Henry S. Contsable

Observations Suggested by the Cattle Plague,

about witchcraft, credulity, superstition, parliamentary reform, and other

matters

Henry S. Contsable

Observations Suggested by the Cattle Plague,
about witchcraft, credulity, superstition, parliamentary reform, and other matters

ISBN/EAN: 9783337152017

Printed in Europe, USA, Canada, Australia, Japan

Cover: Foto ©berggeist007 / pixelio.de

More available books at **www.hansebooks.com**

OBSERVATIONS

SUGGESTED BY

THE CATTLE PLAGUE,

ABOUT

WITCHCRAFT, CREDULITY,

SUPERSTITION,

PARLIAMENTARY REFORM,

AND OTHER MATTERS.

BY

H. STRICKLAND CONSTABLE,

WASSAND, HULL.

" I have read your Pamphlet with much interest. People believe in
contagion as they used to believe in witchcraft, and as fools now
believe in spirit rapping."

Extract from a Letter written by Sir George Cholmley.

LONDON :

DALTON & LUCY, BOOKSELLERS TO THE QUEEN,

AND TO H.R.H. THE PRINCE OF WALES,

28, COCKSPUR STREET, CHARING CROSS.

1866.

CONTENTS.

CHAPTER I.

DURING the last year we have all heard and read of many valuable precautionary measures that have been made use of with great success against the Cattle Plague, such as hanging camphor bags or strings of onions round the necks of all healthy animals, painting their noses with tar, &c.; but I have been much surprised at the ignorance that has been shown of measures that have been made use of in former times for similar purposes. In fact, no mention, as far as I have seen, has been made of any one of them. For instance, there is the word Abracadabra. Serenus Sammonicus tells us that he found the use of this word of the very utmost value in all cases of tertian fever. No doubt Rinderpest is not tertian fever. Still it is a fever. So Abracadabra ought at least to have been tried. We are told that the word has no influence unless it is written in the following manner, and then either attached to the patient or else fixed up in some immediately contiguous locality :—

Abracadabra

Bracadabr

racadab

acada

cad

a

Then there is Camillus Leonardus. No more notice has been taken of what he tells us, than if he had never existed. I suppose the fact is, we are all so full of our Reform bills, spirit rappings, railway schemes, Davenport Brothers, and other interests of the day, that we have no time or attention left for the valuable experiences and discoveries of by-gone times. Now let us see what Camillus Leonardus tells us. He says the real remedies for all diseases are *the precious stones.* He says that the emerald prevents epilepsy, and unmasks the delusions of the devil. He says the serpentine cures dropsy, because if people stand with it in a very hot sun for three hours, they break out into a profuse perspiration. He says that red coral strengthens digestion, if worn about the person; that red cornelian cures dysentery; that green jasper *prevents fever;* and that chrysolite held in the hand *cures fever.* Now the cattle plague is a fever. No doubt a cow cannot hold a precious stone in its hand, but it can wear one about its person; and what I say is that these things ought at least to have been tried.

And now I will proceed to mention one or two rather curious historical parallels.

In the "Hexenhammer," a book of witchcraft, published first in 1489, we read that the devil teaches witches how to make magic ointment for the destruction of cattle, and that if the door posts of cow-houses be smeared with this ointment, the cows will immediately become diseased.

In the same book we read how a witch gets milk from any cow however distant. She sticks a knife into the wall, takes a pail between her knees, and cries to the devil to send the milk of any cow she wishes; upon this the milk immediately runs down the handle of the knife into the pail.

In 1669 sixty-two women in Sweden were condemned to death for bewitching a great number of children. They confessed that they had been in the habit of taking children to a rock in the sea, called Blokula, where they met the devil. He was usually dressed in a grey coat, red breeches, and red stockings. They rode through the air to the rock, with the children on goats. The way was to run a long spit through the goat; then the witch rode on the goat, and the children rode on the spit behind. When they got to Blokula, the devil feasted them on cabbage and oatmeal porridge. The younger witches were not admitted to the feast; but were set to take care of the toads, and keep them in order with long white sticks. The witch's husband, who all the time has a piece of wood in bed with him, in the place of his wife, knows nothing of the affair. Then there were dances. When the devil was in a cheerful humour, his amusement was to make the witches ride on long poles, and then all at once he would pluck away the poles, when down would tumble the witches, to his very great glee. All these sixty-two witches were executed.

In 1303, an English Bishop was accused of evil

practices, and especially, " quod diobolo homagium fecerat et eum fuerit osculatus in tergo." The same accusation was also often made against witches.

Zimmerman tells us that there were in his day devout persons who by earnestly transposing their self-consciousness to the pits of their stomachs, had the power of merging themselves into the absolute divinity, which appeared to them as a pure white light. For some days, they lived upon nothing but bread and water ; sunk into deep silence, turned their eyes in deep concentration of soul to the point of the nose and then the white light appeared.

Khan Helmont tells us that some men have the power of killing animals by merely looking at them. Rousseau says of himself that he had often killed toads in this way. But that one day when he was trying it, the toad, finding he could not escape from his eye, turned round, blew himself up and stared at him so fiercely that he fainted, and was thought for some time to be dead ; but at last he was brought round by treacle and the powder of vipers.

In Sprengel's " History of Medicine," we read as follows :—" Out of every body proceed indivisible " substances and emanations, which diffuse them- " selves through infinite space. Therefore, bodies " can operate on others at any distance, and a man " can impart his thoughts to another who is hun- " dreds of miles off. Also between all bodies there " is either positive or negative magnetism. In the " one case the emanations from each one pass

" onward to the other; in the other case they are
" broken and thrown back, thus causing antipathy.
" It is the same in the vegetable world. Thus, for
" instance, the vine has a decided antipathy to the
" cabbage, &c."

In old medical works we have read that epidemics and plagues are caused by poisonous vapours
arising out of fissures in the crust of the earth.

In the newspapers printed in the winter of the
year 1866, we read of a vast destruction amongst
cattle, caused by poisonous influences which exist
without beginning in, and flow without cessation
from, some remote but unknown corner of the
Russian deserts. We read of mysterious emanations that were carried by the winds for miles, and
left death wherever they lighted. We read that this
subtile power was conveyed by the birds of the air,
by insects, by mice, by cats, by dogs, by rats; and
that it was even carried and imparted by the very
men who were appointed by the Government of the
day to prevent its spreading.

In the newspapers of the same year we read of a
respectable inn-keeper, who wishing to discover a
theft, suspended a Bible to a string, and then set it
spinning. The position of the book when it ceased
to revolve indicated the locality of the stolen property, and the guilty person was discovered.

In the newspapers and other publications of
1866, and of many previous years, we read that
there were certain persons who had the power by

their mere presence of making heavy tables stand for a considerable time upon one leg, at an angle of 45°, and that without in the slightest degree interfering with the equilibrium of the objects placed upon it. Even high lamps remained unmoved, and tumblers full of water preserved their equilibrium without overflowing.

The way the same class of ideas and beliefs take different forms in different centuries, seems to me very curious.

Again, here is another curious illustration of the way history repeats itself.

In the introduction to the Hexenhammer, we read:—" Even preachers of the Divine word were " found who did not hesitate to assure the people " that there were no such things as witches, and " that they had no arts by which they could injure " men and animals; by which imprudent language " the secular arm was not unfrequently restrained " from punishing such sorceries, and thus they " became amazingly increased, and heresy became " enormously strengthened."

So in 1866 we read in the newspapers of the period, that even magistrates were found who did not hesitate to assure the people that the prevailing ideas about the cattle plague were absurd, exaggerated, and superstitious; by which imprudent language the precautionary measures became in those localities less stringent, and the plague raged with increased violence.

Mr. Gamgee has written a book which he entitles, "The Cattle Plague." The first part I shall mention presently. The second part is devoted to an account of the proceedings of a Grand International Veterinary Congress, that seems to have been held at Hamburg and Vienna. It is exceedingly entertaining. Such a charming farrago of antiquated ideas about contagion, infection, Government interference, official inspections, police regulations, veterinary supervision, restrictions of traffic, quarantine, inoculation, &c. &c., I could not have believed possible to find extant in any civilized country in this the nineteenth century. I laughed a little at the report of our own Commissioners on the Cattle Plague, in consequence of its inconclusiveness, want of unanimity and strong veterinary flavour, but it was the wisdom of Solomon in comparison with these German enunciations. And yet I would not for the world say a word against the Congress. It seems to have been a very pleasant festive gathering, if it was nothing else, and the members of it seem to have displayed throughout much heavy German good humour and geniality. I can fancy I see them—hazy figures looming through the tobacco smoke, with their long yellow hair, long yellow pipes and heavy rings on forefingers, neatly finished off with artistically-executed Indian ink sort of mouldings towards the extremities. The difference of opinion upon every subject seems to have been unbounded. But every now and then the question was put whether the respective Governments should or should not be recommended to appoint competent veterinary surgeons to be paid by the State for such and such a purpose, and whenever this was done the general unanimity and harmony that immediately prevailed was such that it must have been delightful to witness.

First, came the question of Quarantine.

Professor Röll thought eight days enough.

Professor Undritz said that cattle from the Steppes sometimes transferred the disease after many months without ever being diseased themselves.

President Hering thought that were the quarantine period to be regulated according to these accounts, its duration might be endless.

Professor Gerlach thought cattle ought to be absolutely prohibited from coming from Russia into Germany.

Professor Haubner thought there should be no prohibition whatever.

Professor Müller thought it no matter whether there was prohibition or not. For that it was always evaded. Sentries had been placed every fifty paces along a frontier, and all to no use.

Professor Hering requested that a decision on the question be arrived at.

Dr. Haubner thought it ought not to be settled by vote, as the result of such a proceeding could not be a definite one. He also said all restrictions were evaded. For instance, Bavaria forbids the import of swine from Russian steppes ; but Saxony does not do so ; so all Bavaria has to do is to receive the Russian swine through Saxony.

Dr. Roll did not believe the cattle plague was communicated by swine, inasmuch as they (the swine) were washed when they crossed the frontier.

Dr. Herting thought the air carries the contagium of cattle plague.

Dr. Müller thought not, or how could places in the midst of infected districts escape as they do.

Professor Hering thought that when scientific veterinary surgeons enunciated their experiences and opinions, the results ought not to be met with incredulity. (Cheers.)

Professor Gerlach acknowledged gratefully the debt due to those who devoted their time to these investigations; but thought it unnecessary through mere gratitude to alter an already formed opinion.

Then came the question of Pleuro-pneumonia.

Professor Gamgee (from England) said, that Pleuro-pneumonia was a disease never taken spontaneously; but introduced into England from the Continent.

Professor Nichlas said, many competent judges were of a different opinion.

Professor Gamgee said, inoculation in this disease was of no use.

Mr. Wüth assured the meeting that, on the contrary, inoculation was of infinite advantage.

M. Marcus thought the pole-axe the only means of combating the disease.

Professor Gerlach thought this measure barbarous.

Professor Nichlas was not for the axe; but for inoculation.

Professor Schell was for recommending to the several Governments compulsory inoculation and compulsory slaughter.

Dr. Haubner expressed doubts how compulsory inoculation and compulsory slaughter could be simultaneously recommended. He was against inoculation at all.

Professor Hertwig agreed with Haubner, partly because he had known a whole herd destroyed by inoculation, which is in fact only intentional infection.

Dr. Fürstenberg stated, that when the Prussian government had rendered inoculation compulsory, he himself had carried it out, and the result was that fifty out of sixty-five died. This fact tended, he thought, to show the non-success in some cases of inoculation.

Herr Schell thought slaughtering should be extended to all suspicious animals.

President Hering thought not, for in some cases all the cattle in the country are suspicious.

Professor Fuchs strongly advocated veterinary surgeons being appointed in all districts, and paid by the State. (Protracted cheers.)

Dr. Müller wished the meeting to come to a unanimous decision about pleuro-pneumonia.

Professor Gamgee said he had obtained great success from the use of sulphate of iron, and he asked whether this remedy had been tried in Germany.

Herr Kaltschmidt said he had been witness where a proprietor had used this remedy perseveringly, but he regretted to say without success, inasmuch as ou of a large herd not one survived.

Professor Fuchs knew an exactly similar instance.

Fourth sitting, 17th July.

Herr Falke thought erysipelas in pigs should be under Government supervision.

Professor Zangger thought this disease did not call for the interference of Government authority.

Professor Gerlach said that in Hanover pigs often had erysipelas.

Then came a discussion on scab. The result of which was, that the whole meeting agreed unanimously that scab should be under State supervision.

The meeting then proceeded to discuss the question, whether the malignant disease of stallions should be placed under Government supervision.

Dr. Röll thought decidedly that it should. The disease had been fully described in the Austrian Veterinary Review. It attacked the nerves, and was attended with dropsical appearances, &c.

Dr. Fürstenberg entertained doubts about this matter, inasmuch as he did not believe there was such a disease.

Dr. Hertwig entertained little doubts as to the existence of the disease.

Finally, it was put to the vote, when it was decided by a large majority, that the disease* should be added to the list of those which were to be placed under State supervision.

Then came the question, whether Catarrhal fever, also called Head disease (in common parlance, cold in the head), ought to be placed under Government super-inspection, then hydatid disease of the brain in sheep, then measles in pigs, &c. At last, the first International Congress of Veterinary Surgeons finished up with a grand, and I regret to say, lengthy discussion upon sheep pox.

In August, 1865, took place a second Congress at Vienna.

The first question discussed was Hydrophobia and the keeping of dogs.

Dr. Von Helm said, that with regard to the keeping ofdogs, the Municipal Council had been engaged during several years devising measures, and would some time ago have come to a decision ; but it was anxious not to exact too much on the one hand, and also from the great importance of the subject not to treat it even apparently superficially, &c.

Then came a long discussion about dogs.

The next question was Rinderpest.

The President proposed that Councillor of State Unterberg, Land Veterinarian Werner and Medical

* I need hardly say that there is no such disease.

Councillor Haubner, be appointed to act on the Committee of Investigation. He then requested those gentlemen who had observed any mistake in the writing of their titles or the spelling of their names, to communicate it to the stenographic officials—(how like a set of ceremonious title-loving Germans).

Dr. Bleiweis thought rinderpest, sheep pest and goat pest identical diseases.

Professor Reynal asserted that rinderpest is developed in ships in consequence of pressing together many animals.

Herr Rawitsch agreed with him, and also that it is specially induced by long marches and privations.

Prof. Haubner thought, judging from his experience, that if sentries were placed along the frontier so close that they could shake hands, it would be no use. Disease cannot be kept back.

Mr. Ernes (from England) said the disease was introduced into London by a cargo from Revel.

Prof. Unterberger said that the disease was already in London before the cargo started from Revel. That, furthermore, there was no rinderpest in or near Revel.

Prof. Jessen said, " with regard to the assertion in England, that diseased animals were taken into Revel from the interior by railway, I take the liberty just to mention that there is no railway into Revel."

Then the committee proposed, that it was desirable that veterinary surgeons be appointed and paid by the Governments for the frontier districts. (Cries of very good.)

After this came a long discussion about the keeping railway waggons clean, at the end of which, the President said, "the washing of railway carriages renders the supervision of veterinary surgeons to be paid by the State necessary; those who agree to this will rise from their seats."

All rise from their seats.

Then the President proposed that they should hold a meeting next Sunday morning, in order to discuss the questions of Hydrophobia and Police regulations relating to dogs.

Next Sunday morning the meeting is held accordingly.

First, the reporter read the report of the Hydrophobia Committee, at the close of which, the Committee recommended:

(1.) An accurate dog census.

(2.) That the identity of each dog be determined by a certain mark or token; the manner of putting this into practice to be left to the local authorities.

(3.) That with a view to diminishing the number of dogs in the country a dog tax be recommended. And for the more effectual arriving at this result, that the tax on the female dog be still higher than that on the male dog.

(4.) That no dog be permitted to go about without supervision.

(5.) That the muzzling of dogs be regulated by competent officers.

(6.) That all dogs suspected of hydrophobia be immediately reported to the authorities.

(7.) That dogs suspected of hydrophobia be kept in secure custody till their state of health has been determined. Those that prove rabid to be killed. The others to be returned to their proprietors.

(8.) The Committee is of opinion that if these measures are intended to be effectual, competent veterinary surgeons to be paid by the State, should co-operate in their execution. (Expressions of assent.)

Dr. Fuchs thought that every dog as soon as it gets a proprietor, should be reported to the police. He thought it ought to be known how dogs often come into families. They are frequently brought home by children, especially young puppies. He regretted that the country to which he belonged (Baden) had made the mistake of taxing the female dog less than the male dog.

Dr. Hertwig thought the equalization of tax upon male and female dogs was of the first importance.

Dr. Deisinger said, that in Bavaria the government had done all in its power for the prevention of hydrophobia, even the smallest dogs had been muzzled, and all to no purpose. He thought that when it is considered how often dogs excluded from their homes at night disturb peaceable people; how often children are knocked down by dogs, and how dogs run about in public outraging every feeling of propriety and good order; when all these things came to be considered, he thought that the strictest code of government regulation would be welcome to the public.

Dr. Gerlach desired, on philanthropic grounds, a law of Draconic severity, the dog being a domestic animal of limited usefulness, but great danger. Although echinococcees may be not common, still they occur, so that it becomes a question, whether dogs ought to be kept at all. Besides which, it is proved that the dog carries the seeds of giddiness in sheep. If dogs were done away with, sheep would no longer be afflicted with giddiness. Dr. Gerlach thought that canine madness should be considered by Government as a purely contagious disease.

Dr. Rawitsch thought nobody would deny that canine madness sometimes originates spontaneously. (Loud cries of "it never appears spontaneously.")*

The reporter thought every dog should be rendered recognizable. (A voice from the centre, "could it not be done by means of a collar.")

Dr. Hertwig was in favour of a tax on dogs. The most faithful dogs will run away when they go mad. In a few days perhaps they return. But what have they been doing in the mean time? How many people have they been biting?

Dr. De Souza (from Portugal) said, that in his

---

* I am astonished at Dr. Rawitsch's courage; considering whom he was addressing, I wonder he was not torn in pieces.

country there was no tax on dogs; yet, notwithstanding that, dogs never go mad.

Dr. Zaugger thought that might be, and yet prove nothing. Madness might not exist in a country for a long time, and then it might break out. At least it was so with hailstorms. Sometimes there would be no hailstorms in a country for a long time, and then a great many would take place. Still he was against a tax. A tax would have little effect. If one man discontinued to keep a dog, another might keep two, and then where would be the good of the tax.

M. Reynal (from France) condemned all muzzles on principle.

Dr. Bleiweis was for rendering it compulsory for all dogs to be led in a cord or chain, for then they could not run about and bite people.

The reporter thought this would lead to great danger to the public. If dogs are led about with cords, how often will children fall over them or become entangled in the string.

Dr. Gerlach thought that, as canine madness arises from bites of dogs, therefore if we prevent the bites we prevent the canine madness. Therefore he was in favour of muzzles.

Dr. Hertwig agreed that the great thing was to diminish the bites of dogs. It is sometimes objected that dogs dislike being muzzled. But if so it is only for a few days. "I have two dogs. When the " muzzle was first put on, what scratching and

" brushing and whining, but now they are used to
" it; and I need only show them the muzzle and
" they come running up to have it put on. I will
" not enter more deeply into the matter. I am in
" favour of muzzles."

Dr. Jessen said he had seen dogs led by a string.
He had seen them led by a child. The dogs often
broke the line and threw the child down, and then
the biting was not prevented.

M. Weber had seen several cases where a desire
to bite was created in dogs by the muzzle. " I myself
" have been bitten in the hand by a dog that had
" on a muzzle. This I consider sufficient proof that
" muzzles do not always prevent biting. I, there-
" fore, am strongly against muzzles."

Professor Hertwig thought a suspected dog should
be kept under police inspection for at least twelve
weeks, for he had known in Berlin a dog discharged
and still go mad.

Next came the question of Laws of Warranty.

M. Schall thought there was great difficulty in the
question, in consequence of the endless difference of
opinion that prevailed.

M. Reynal thought an international law of warranty could not be framed, because diseases differ so
much in different countries.

Director Gerlach, on the contrary, thought staggers in France would be staggers in Germany, like-
wise asthma. He also protested against the con-
temptuous treatment of the veterinary profession
that was so general.

23

Then came the question of warranty for crib-biting.

Dr. Zlamal said crib-biting was very detrimental, because all the other horses in the stable with a crib-biter contract the same habit.*

Dr. Stichel said, that in Russia military horses were fed by nose-bags, and still they were crib-biters.

Dr. Zlamal thought this observation deserved great consideration.

After this came the question of cataract.

Dr. Zlamal was decidedly in favour of warranty for black cataract but not for grey. Of the green cataract he did not intend to speak.

Dr. Pillwax believed it to be an established fact, that a horse which is half blind cannot altogether be depended on.

Dr. Siefman agreed with Dr. Pillwax, that objects which a horse could not distinctly see might frighten it.

Dr. Frank thought the law of warranty ought to extend to green cataract.

Nothing was said about cataract of any other colour.

The next subject was ringworm.

Dr. Gerlach protested against the expression " ringworm."

* Contagion again. Of course this is the old delusion, as in the case of almost all these supposed contagious diseases. The same bad management produces in all the horses the same crib-biting habits, and the first that began it gets the credit of giving it to the others.

The reporter said it was the right one.

The President begged the gentlemen in favour of the word "ringworm" to rise.

They rise.

The expression ringworm is carried.

The next question was, whether there should be a law of warranty about trichinæ in pigs; that is, whether there should not be a law that the purchaser of a pig might; on the discovery of trichinæ, drive it back to its former owner, and receive again his money.

Dr. Zlamal thought there was a difficulty about this question, inasmuch as it was impossible to detect the presence of the disease till after the death of the pig, and then only with a microscope : therefore it seemed to him that a law of warranty would be nugatory.

Notwithstanding this objection, the question was carried in favour of the law by a large majority.

Finally, the President wished statements to be made respecting the effect their deliberations at the previous Congress had had upon the various Governments.

Dr. Hertwig was very sorry to say that they had as yet borne no visible fruit in Prussia. The decisions had had no visible influence whatever upon the legislation.

Dr. Müller said the same with regard to Austria.

Dr. Fuchs thought the efficient organisation of the veterinary practice of international importance.

Dr. Iwerson (from Holstein) regretted that in his

country no legislative measures had as yet been enacted, but we live (he said) in hopes.

Dr. Straub (from Würtemberg) was gratified to say, that the Government of his country had issued an order that railway cars should be kept clean.

I think, perhaps, I have extracted enough just to give an idea of what fills more than three hundred mortal pages of Mr. Gamgee's book. And now what are we to say about all these good folks? There can be no doubt what we should call them in England. It is some time since I set to work to forget my German, and yet I fancy I am right in thinking the German equivalent of the English idiomatic expression I am alluding to must be " Alte Frauen." Still I may be mistaken.

Amongst such a mass of evidence and opinions, twaddle as it must seem to most Englishmen, there cannot but be something worth noting.

In the first place then, the prevailing idea is that the Rinderpest comes only from Russian cattle. But the only way they can get over the innumerable cases that apparently contradict this, is by saying, which they mostly seem to do, that Russian cattle will impart the disease in any country, at any distance of time after leaving Russia, and without ever having any disease themselves. Allowing this as an argument, of course they can prove anything.

But Professor Jessen, the Russian veterinary representative, *denied that the disease does originate in Russia;* and each representative denied that it

originates in his country. So we come to this, that
the original disease is only to be found, like Mr.
Grimes in " the Water Babies," at the other end
of nowhere. Well, it is a very old story. Every
country wants to shift the balance on to its neigh-
bour's shoulders, like a set of school boys, " Please,
" sir, I didn't do it, sir; he did it, sir, please, sir."

Again, in Oliver Twist, " Yer did it, Charlotte,"
said Noah Claypole, when they were detected after
robbing a till. " Yer did it, Charlotte; yer know
" yer did."

But after all, the best explanation of this tendency amongst half-cultivated people to deny spontaneous disease, is this. A. has the cholera. Doctor
B. comes and says, A. caught it from C. In this
way he escapes, having to tell the truth, which is,
that he does not understand why A. has got the
cholera; for to say, " I don't understand," is a
thing the human soul as a rule abhors. But even
supposing A. has caught the malady from C., the
mystery is only removed on to the next peg; there
it is yet. Still it is taken away from Doctor B.'s
eyes, and he is satisfied. But sending a puzzle
into the next room does not solve it. It is the same
with some of the shallower men of science. They
remove the first cause a long way back, and then
think they understand everything, when in fact they
understand no more than they did before; or not
so much, for probably before, they at any rate
understood that they understood nothing. So again

with their anxiety to deny the possibility of spontaneous generation. They think they understand more about it if they shift it back a more or less long time, but it is all a delusion. Besides, what is time?

Does not Tennyson say : —

> " For was, and is, and has been, are but is ;
> And all creation is one act, at once
> The birth of light: but we that are not all,
> As parts can see but parts, now this, now that,
> And live perforce from thought to thought, and make
> One act a phantom of succession ; thus
> Our weakness somehow shapes the shadow time."

And does not Archbishop Whately say that our notions about time are delusive; arising as they do only in consequence of the succession of our ideas. There is an analogous propensity amongst critics ; namely, a perpetual striving to deny originality, and to accuse of plagiarism. The fact is, they well know that the few ideas that are in their own heads were only got there through much tribulation from other sources, so they cannot believe in ideas ever coming in any other way. For people proverbially judge others by themselves. The melancholy nature takes sombre views of people and things ; the cheerful take cheerful views. The poet who has a sense of beauty dwells on the beauty to be found in all things, whilst the mind that is dead to beauty sees none. Pythagoras and Jedidiah Buxton had preternatural souls for number, so the former

thought number the ultimate principle of the universe; and the latter liked going to a play, because
it amused him to count the words the actors
spoke, and the number of steps of the dancers;
but he did not like Garrick, because he spoke so
quick he could not count the words; and he
would answer in a second an arithmetical question,
that would take most people a quarter of an hour
to calculate. Analyzing minds can only see distinc-
tions in things; synthesizing minds can only see
resemblances. So, as I say, the mere critical and
negative mind can see nothing in the universe but
imitation, repetition, and non-spontaneity.

It seems generally thought that the question of
contagion or non-contagion in any case is very easy
of proof; but it is on the contrary almost impossible
to prove one way or the other.

A boy at a school has the measles. Now boys
at school live such similar lives, and are subject to
such similar influences, that I should naturally
expect that the same atmospheric and other conditions that gave it to one boy would give it to
many others. And yet when they do take it,
everybody concludes at once that they all caught
the complaint from the first one that had it.

No doubt absolute proof is impossible one way or
another. Still there is one thing quite certain, that
every disease has originated spontaneously at some-
time; so in the question whether a disease, say
small-pox, is or is not ever taken spontaneously;

the only absolute certainty is on the side of the former.

Mr. Gamgee says, as we all know, that this cattle plague is never taken spontaneously. I say it is. I say I know half-a-dozen instances within five miles of my own house, some of them my own tenants, where there had been no communication, and therefore no contagion, and yet the plague broke out. Oh, but (our Professor would say) the air carried it, or a dog, or a cat, or a bird, or a rat. But if this is called an argument, I will prove the same about anything whatever. Broken knees in horses, for instance. Suppose I choose to assert, that broken knees can only be taken by contagion, Mr. Gamgee would doubtless bring several instances to show the contrary. I say these instances prove nothing; because the air will carry "tumbling down" in horses, any distance almost, and insects carry it, and dogs are known by everybody to carry it constantly. The proof is absolutely as complete in the one case as the other.

In one place Mr. Gamgee tells us how forty cows in Lancashire died of the plague, and how they were known to have caught it by contagion, because a short time before it broke out, a box of butter had arrived from an infected district at the farm where they were.

The fact is, the only real argument Mr. Gamgee has that will bear looking at, is the assertion, no doubt true enough, that one beast will often catch the complaint directly from another; but to call this a proof that it is never taken spontaneously as well, is like showing that it often rains on Wednesday, and then calling that a proof that it never rains on Thursday.

What I say is, that as in innumerable cases the disease seems to be taken spontaneously, the *onus probandi* rests with those who assert that it cannot be so taken.

Why one village is attacked and the next escapes no one in the present state of science can say, any more than they can in the case of cholera, or any other epidemic. Why will there be at any time many cases of a common cold in one village, and none in the next? From what differences in the conditions of the atmosphere is it that colds will take one form one season, another form another. This last winter and spring they have very much taken the forms of ear-ache and swelled glands. Why is all this? The only answer is, we cannot tell. But this is just the confession, human nature in its ignorance, pride, and vanity, hates to make. So then come the idle theories, as for instance, in former times that Jews poisoned the wells, which accounted for the black death; or that subterranean vapours issuing from clefts in the earth's surface, caused epidemics; whilst at the present time we all talk about mysterious contagious emanations.

CHAPTER II.

ERRONEOUS POPULAR NOTIONS ABOUT DISEASE.

LAST winter I published a pamphlet about the
Cattle Plague. I endeavoured in it to show that
most of the popular ideas about diseases and contagion are either wrong or exaggerated ; that Miss
Nightingale was right when she ridiculed the notion
of diseases being so many separate positive entities,
only to be reproduced from themselves, as if they
were organized beings like a cat or a dog ; and that
dirt, vice, or the transgression of nature's intentions
will produce any disease ; that every cause must have
its effect; that every departure from right in this
world must have its consequence, and must be paid
for either in person or vicariously ; that diseases cannot be separated by hard decided lines as usually
supposed ; but that they dovetail into each other with
infinite and indefinite complications—like colours;
that although there are bay horses and brown horses
there are also many that are as much one as the
other, whilst sometimes a horse has to be described in
racing entries as either grey or roan, or black or
chesnut or brown ; that is to say, a complication of
all colours. I pointed out that the majority of farmers
in winter keep their cattle in an unhealthy manner :
those for the butcher overfed, the others underfed,

and all without natural air and exercise; that amongst
men such modes of life have always ended in destructive typhus diseases ; that malignant goal fevers
resulted formerly from the horrible state of prisons ;
but that filthy as they were, they cannot possibly
have been so filthy as the average cow-shed of the
country; that being so kept, it was little wonder
that in most years pleuro-pneumonia and other forms
of internal inflammation should be so fatal as they
are, or that once or so in a hundred years when an
extraordinarily unhealthy season occurs typhus in
its most malignant and contagious forms should
break out. I showed that cattle doctors of the present day are in knowledge where man doctors were
a hundred years ago; that their treatment of sick
cattle does not give them a chance ; and that if in
a bad case of low fever, such as Rinderpest, a beast
be shut up in the dark in a temperature above 40°, it
requires a miracle to save it; and yet that this is the
universal mode of treatment by the whole profession ; and I showed how, in the very few cases where
nature and open field treatment had by some acci-
dent been resorted to, far the majority had recovered. Furthermore, I mentioned my conviction,
that an animal that has been perfectly managed, that
is to say, an animal that is in the highest possible
state of health and vitality cannot take any disease
whatever whilst in the open air, and therefore that
the simple and certain way to prevent these diseases
in future, is good management; that is to say, the

best of food, the purest of air, no matter how cold, and natural exercise in the open fields, day and night, winter and summer; or, if they must be shut up, that they should only be protected from wind and wet; but not from one ray of light or one degree of cold. Animals kept in this way never catch a cold, if they are well fed ; and therefore never have any of these internal diseases, for a cold is the beginning of them all. I have been told that the greatest sufferer in Cheshire has been a nobleman whose cowhouses were celebrated for their perfection; and this is just what I should have expected. I have all along called the disease " cow-house typhus," corresponding to jail fever amongst men. No doubt it may also in innumerable cases be called " famine typhus ;" but that only means that as amongst men so amongst cattle, impure air, crowding together and insufficient food will, any one of them, cause typhus diseases.

I said in the pamphlet I published that I doubted whether the Commissioners who gave us such long reports knew much about the habits of animals, or even such a simple fact as their never, unless forced, going under cover for shelter from any weather whatever, though they will do so to escape flies in the heat of summer.

I have six Alderney cows, the most tender of all kinds. Ever since last November they have been out in a field day and night. It is now the end of May. During all this time they have been in perfect condition, and giving as much milk as ever.

They have a comfortable open shed to go into if they like. My cowman told me the other day that not one of them has ever gone into it the whole winter. It has been a remarkably wet, stormy winter, and the thermometer has some nights been at sixteen degrees below freezing point. The Plague has been all round me. The way I do is this. I go and spend half an hour in the sheds of some farmer whose beasts are dying all round me. Then I come home, and immediately visit my cows in the field. They are very tame; so I first look which way the wind is blowing, and then I go to the windward side of one of the cows, so that none of the emanations may be lost, and stroke its head for a considerable period. I have not succeeded hitherto in imparting the disease, and I am sadly afraid my endeavours to do so would only lead to disappointment, however long I might continue them, or however often I might repeat them. Now I know perfectly well what Messrs. Gamgee and Co. would say here, for they have said it before in similar cases. They would say, " The cows are Alderney cows, and therefore all this proves nothing." It is like the elder Mr. Weller, when his son consults him how he should end his valentine. " I could end with a werse," he says, " what do you think ?"

" I don't like it, Sam," answered Mr. Weller. " I never know'd a respectable coachman as wrote poetry 'cept one, and he was only a Cambervell man, so that proves nothin."

Furthermore, I pointed out how people are apt to be deceived by words that are only used for colloquial convenience, such as "disease," "cholera," "enteritis," "Rinderpest," &c. ; that the habit of using these and other terms lead many who do not think immoderately to suppose they have positive meanings ; that different (so called) diseases, instead of meaning only different organs out of order in different ways and places, mean so many different entities ; that, for instance, "enteritis" means an actual something, when in fact it is not a something, but only an inflamed condition of a something, that is to say, of the duodenum, whatever the duodenum may be, for I have not the least notion. Of course this is plain enough when pointed out, but still people generally do not seem to regard it in this light. Probably all these names have had for their origin the endeavour to hide ignorance, and this would account for our giving them in these cases, but not to the different internal disorders of a watch or a steam-engine, for these latter we can take to pieces, understand, mend and put together again ; but if we could not do so, and a watch went badly, I have no doubt whatever that we should say, it had got that terrible disease, "horologiumpest." This would at once satisfy us like children when told the mere name of a thing, and prevent our having to say we do not understand what is wrong ; and at the same time it would (a thing the human mind craves) set our imaginations, our credulity, and superstitions free to theorise about it and say, that "horologiumpest" was a fearful disorder to which watches are subject; that one watch never takes it except by contagion from another ; that the salubrity of the English climate is such, that the disease is never taken in this country spontaneously, but that it is always imported direct from Geneva. The proof of this being that watches in Geneva are notoriously known to be often afflicted with the disease, and that large numbers are always being imported into England; furthermore we should say, that keeping the watch clean and treating it well, though desirable things enough in themselves can do nothing towards preventing this terrible complaint, and that the only real measure is, immediately a watch is observed to go wrong, to stamp upon it until it has ceased to tick, and then bury the fragments in the ground, the chain and key to be either buried with it or carefully disinfected with carbolic acid.

"What is this funny little round thing, papa ?"

"It is a 'pillula calomelanos et opii,' my dear." And away the little boy runs perfectly satisfied ; his fancy setting him to work with all sorts of credulous imaginings about his newly-acquired "knowledge." Are we not all like children ?

CHAPTER III.

ILLUSTRATIVE ANECDOTES.

THE time for talking about treatment of Rinderpest is no doubt past. The act for the prevention of curing sick cattle has rendered all such discussion useless. Still a time may come when it will not be illegal for a man to cure his cow when she catches a cold. So I will say a few words upon the subject.

Nature, good food, broad daylight as long as it lasts, open fields and open air, no matter how cold. This is the only treatment for sick animals that is worth two straws; and this is the only treatment that never has been practised except in a very few accidental cases. And in those few cases far the majority recovered.

The following is an extract from a letter I received from a relation of mine last January.

"There has been such a curious illustration of the wisdom of keeping beasts cool in the plague, that I have been wishing, day after day, to tell you about it.

"John Farrow, of Alkbro', in Lincolnshire, after having lost every one of a lot he had shut up and kept warm, found another lot beginning to fall ill of the complaint. In moving them from one field to another, one fell into a drain and got considerably chilled, but was observed to go on better from that time. Observing this, and being forbidden by the Inspector to bring the animals along the road up to his buildings, J. Farrow determined to try cool treatment, and with that view, made a sort of slight cover for the beasts with a piece of old rick cloth, just

D

to keep off rain; and then he made a low heap of straw like a wall, perhaps three feet high, but between the straw and the rick cloth was a space of perhaps two feet all round, so that the wind blew through in every direction. This was in one of the bleakest fields on the whole estate. The result was that at the end of six weeks, out of twenty that had fallen ill, only three had died, though some had been so ill that they did not seem able to stand, and as J. Farrow himself expressed it, were worse than those which died. The last time I saw him was after the bitter cold of the 11th and 12th of this month,* when the frost, and snow, and wind that came suddenly was enough to try anything that was easily damaged by a low temperature. I asked him particularly, how they had borne it? and he said, they were not a bit the worse and were getting well without any check."

I have been told that some diseased beasts belonging to Lord Petre in Essex were tied to trees in the open field, and that they all recovered.

Last December a farmer near Malton employed a man to kill and bury two beasts that seemed in the very last and worst stage. But the work was hard, and the day was wet, so when half the task was done the man went into a neighbouring public house to refresh himself. By a most happy chance he got very drunk, and remained so for four-and-twenty hours; then when he returned to his work he found the remaining beast quietly grazing by the grave of its companion, and the next day it was quite well. The fact is, the two animals had all but been killed in the orthodox manner by heat and darkness, and the cold pure air of a frosty December night acted like a charm.

* January.

In January, at a place called Whitby in Cheshire, some diseased beasts were turned out in a field where there was some water at the bottom of an old lime-pit. They all recovered. Then there was a rare to do. Some wise men posted off to analyse the water. Some wrote to newspapers to say they had been in the habit of using lime-water all their lives for all sorts of diseases, and never lost a single patient. Some wrote to say they were perfectly astonished that such a well-known remedy had not been tried before ; and so they went on for about a fortnight, till at last the discovery was made that lime-water was not a bit more use than arsenicum, Turkish baths, or ferruginous preparations. Of course, the truth is, the animals recovered because they were turned out in the field and left to nature. In the most malignant cases there is no hope, nature herself can do nothing, but such cases are exceptional.

Not long ago I knew of a supposed hopeless case of low fever. Dr. Baker Brown was sent for. He came, opened all the windows, pulled the bed to them, and let into the room all the light, air, and sun he could, and the patient recovered. The neighbouring surgeon, as I was told, said this was all very well for a man with a great reputation, but that if he had done it he must have lost his practice for ever, such being the popular abhorrence of fresh air.

A tenant of mine had forty animals attacked. This was after the Act for the prevention of recovery had come into force, so they were of course shot.

Otherwise the disease was in so mild a form, that no one, the inspector included, doubted that they would mostly have recovered.

In fact, in this degree of it, the complaint is nothing but a feverish influenza cold. One of the beasts had it in the most malignant form. I asked the farmer if it were different at all from the others. He told me it was in the lowest condition of any, and had been shut up all the winter in a dark box.

I knew of a farmer who took every conceivable precaution. He allowed no communication with his cattle. He watched the dogs, he killed the cats, he warned off inspectors, and on his return from market every week he put spurs to his horse, and galloped as hard as he could for a mile, when he got to where an infected farm was situated about a hundred yards from the road. Yet it was all to no purpose, the malady reached his farm, and his loss was great. Of course the explanation is, that though he could warn off inspectors,* and watch the dogs, and tent the birds, and kill the cats, he could take no measures against the north-west wind which blew from the infected farm about ten miles off.

* The belief that drovers and veterinary surgeons carry the disease is explainable in a very simple way. The plague is perpetually breaking out at innumerable farm-houses all over the country. Drovers and veterinary surgeons are perpetually visiting farm-houses all over the country. One of these men goes to one, and next day the plague breaks out. Everybody says he brought it. But not a word is said about the ninety-and nine cases where no disease followed the visit of a drover or cattle-doctor.

CHAPTER IV.

DR. PRUNER in his work on the diseases of the
East describes the cattle plague in Egypt. According to him the chief causes of it were the use of
improper and insufficient food, and animals being
cooped up and crowded together. Mr. Gamgee's
proof of the fallacy of this opinion is charming.
He says, " no weight can be attached to Dr.
Pruner's theory as to the cause of the disease, from
the fact that similar views are constantly entertained
regarding fresh outbreaks of the plague." I suppose
he would say that no weight can be attached to the
idea that food allays hunger, from the fact that
similar views are constantly entertained regarding
fresh instances of accessions of hunger.

Mr. Gamgee says Aberdeenshire is an instance
of the success of stamping. out. To be sure the
disease reappeared five times in that county, and had
to be stamped out five times ; still this only shows
how useless is any system limited to one county.
Now what a delicious muddle this is. The rapid
transition of argument to account for the second fact,
which contradicts the first, is like the school boy and
his penknife.

"I say Dick, lend me your knife."

"I can't, I haven't got one, besides I want it myself."

Surely Mr. Gamgee must be the man Sydney Smith talked about, whose understanding was always getting between his legs and tripping him up. And after all it is quite possible, for Sydney Smith has not been dead so many years.

But this Aberdeenshire stamping out is just like all Mr. Gamgee says in his book about stamping out on the Continent, only he evidently cannot see the true conclusions from his own facts. Some of the countries in the middle of Europe seem always to be stamping out the plague ; and then the plague when it is stamped out seems always to be coming back again. The fact is, the whole thing is a delusion. Of course if every sick beast in a country is killed on the 31st of March, the probability is there will be no sick beasts in the country the 1st of April. So it seems to all, who are not in the habit of thinking, immediately to be stamped out. All such (and they are a good many on the whole) are just now thinking that the stamping out that is now going on accounts for the diminution of cases that has undoubtedly taken place of late. The remaining few remember that according to all history, and all accounts of Rinderpest, the disease always rages worst in winter and diminishes or dies away as the summer approaches. This is specially mentioned in the first report of the Commissioners with regard to the murrain in England a hundred years ago.

I know that just now my ideas about these things must seem to most people crotchets; that is to say, my ideas that stamping out diseases is a delusion; that sick animals should be left to nature, open fresh air and good food, and that the popular ideas about contagion are partly much exaggerated, and partly sheer nonsense. A hundred years ago denial of the existence of witchcraft was considered by all right thinking people, from Addison downwards, a wrongheaded crotchet. Mr. J. S. Mill says, very truly, the crotchet of one age is the acknowledged truth of the next, and the truism of the third.

Between two and three hundred years ago Oriental plague was common in London, breaking out violently at different intervals. I have never read that it was stamped out. But I have read that gradually people adopted cleaner and healthier habits, and as we all know, the plague gradually ceased.

Even up to thirty years ago people used to talk about Glanders in horses, just as they now do about Rinderpest. Stable habits though still bad enough became far less filthy than they were formerly, and such a thing as glanders is never heard of in England now.

Wrong and insufficent feeding, bad air and bad management, each and all get animals into such a state that they are ready for some internal in-

flammation at the slightest change of weather ; sometimes this inflammation flies to the lungs and its
neighbourhood, and then comes pleuro-pneumonia in
its various forms; other seasons it will take typhus
forms. Once in a hundred years the unhealthiness
of the season is excessive, then we have cattle plague
in its present malignant form ; and here is the whole
matter in a nutshell. By the bye, I do wish people
would make a little more use of these nutshells than
they do. Sir Cornwall Lewis said, "life would be
very enjoyable if it were not for its pleasures." The
old port wine drinking Scotch Laird who hated
everything that interrupted his potations said, life
would be very enjoyable if it were not for that bane
of society, "conversation." I say, life would be
very enjoyable if we were not deluged with words as
we all are ; and when the pomposity of pedantry is
added, the burden of them becomes intolerable, unless
one can treat the case after the manner of Archbishop Whately. One day a medical gentleman was
indulging in rather longer words than the Archbishop thought quite becoming ; so he all at once
said, " By the bye, Mr. ——, what causes death in
" hanging ?" "The explanation, your Grace, is
" this : in hanging, inspiration is checked, the
" circulation is stopped, and blood suffuses and con-
" gests the brain." " Bosh !" said the Archbishop ;
" the reason is because the rope is not long enough
" for the man's feet to touch the ground."

Mr. Gamgee's book has 859 pages.

Mr. Gamgee says, that the reason the disease is still with us is that the stamping out measures have not been stringent enough. But we read that in Italy, in 1713, the Pope made a law that all diseased animals were to be immediately slaughtered, and that every body who in any way infringed or impeded the law was to be immediately slaughtered also. Surely that was stringent enough. And yet the plague was not stayed for a whole year, when no doubt a healthy season came and stopped it.

What a strange thing it is that after all our so-called progress we should return, as we are now doing, to the barbarities of the ignorant and superstitious ages.

Mr. Gamgee says cattle plague is absolutely different from small pox. Dr. Murchison says they are very closely allied. Now, with all my respect for the veterinary profession as a body, I must say I cannot feel so much confidence in the opinion of one of its members, as I do in that of one of the first physicians in England.

Oh, but (says Mr. Gamgee) vaccination was no use, so the diseases must be different. He might just as well say hunger in a man is different from hunger in a cow, because roast beef cures one and not the other.

A few months ago Baron Rothschild's cattle were attacked with the Rinderpest. Some of them seemed for a time to be recovering under Mr. Worms' treatment. Oh, but (said Prof. Simonds) it is not

Rinderpest at all. It was very injudicious by the
bye, his being so precipitate, for they all died after
all.

All the English veterinary surgeons said the
plague first came by railway from Russia, through
Revel to Hull. But Dr. Jessen (a Russian) said
there was no plague in or near Revel, and no railway into Revel. Oh, but (said Mr. Ernes), the
Revel cargo did not come from Revel, it came from
Riga. In Aberdeenshire the disease was stamped
out, and re-appeared five times. Oh, but (said Mr.
Gamgee) this was only because it was not stamped
out also in the neighbouring counties. Verily, the
knowledge of these learned professors is so profound
and multifarious, that they have an answer ready
for every thing.

As I have said above, it is quite true that we cannot understand why one district is affected whilst
the next one escapes. There is always the same difficulty whether the question be of rinderpest in cattle,
or cholera, or any other epidemic amongst men, or
even common colds. But because we cannot understand it, that is no reason why we should invent
what we call explanations, and talk about subtile
emanations carried by cats and dogs. Formerly
people often could not understand why a cow was
taken ill, but that was no reason why they should
burn a witch for being the cause of it. Surely if we
do not understand a thing, it would be much better
to say so at once. Luther, when he was once con-

sulted about a sickly child, said it was possessed by
a devil, and he recommended its parents to throw it
into the river and drown it. Physicians in a succeeding
age, for a like case, exhibited rhubarb and magnesia,
and calomel. At last, in the present time, they have
come to open air and good diet. But the veterinary
profession are still where Luther was, for when a cow
is ill they say she is possessed by a fearful influence,
and that she must be slaughtered forthwith, lest
other cattle should become similarly possessed.

Another folly is that of burying and wasting the
hides. The evidence at the Hamburg Congress was
that the hide of the most diseased animal completely
loses all power of infection after being hung up in
the air for twenty-four hours.

Mr. Gamgee is a very clever man. His knowledge of facts connected with his profession is undoubtedly great. He has evidently got what Sydney
Smith calls that tremendous engine of colloquial
oppression, " a good memory." O. W. Holmes, the
American, truly says, " The man to be depended
upon is not the man of facts, but the man who can
think correctly about facts."

The Chinese know facts. They knew the mere
facts about gunpowder many thousand years before
the Europeans did. But they knew nothing more,
till we showed them its enormous value by shooting
them with it. Thus Mr. Gamgee possesses great
power of seeing mere isolated facts and differences,
but none of seeing resemblances, and of discovering

unity of principle. He could distinctly see that a cow has a straight back, and a camel a crooked one, and he would conclude therefrom, that a cow and a camel are two totally different animals. But he could with difficulty be brought to see, that as a cow and a camel have each of them seven neck bones and four stomachs, there is a unity of principle between them. Thus he cannot see the unity of principle between typhus in a man and rinderpest in a cow, because he has only an eye for the differences. He says when you cut the cow open, you find an inflamed membrane here, and an inflamed membrane there, a disorganized appearance about the second stomach, an unusual look about the third, and something or other else about the fourth, all of which, he says, is totally different from typhus in a man. Of course it is. How can a man possibly be disorganized in his third stomach, when he hasn't got one.

I was told the other day of a little boy who had got the mumps. He was hungry as an ogre, but could not open his mouth to eat. So the country apothecary was sent for, who immediately pronounced it not mumps, nor anything to do with mumps, because the swelling began half an inch or so lower down in the jaw than is usual in mumps. And no doubt, if he had been allowed, he would have proceeded to treat it as some mysterious, awful and unknown malady. And this is the way we allow ourselves to be bamboozled by men, who have senses like a dog or a cat, to observe detached

objects, but no brains to compare and connect them
like a reasoning being.

Ruskin says, " Analysis and seeing only dif-
"ferences, tends to vice and deformity, whereas
" synthesis and seeing resemblances, tends to unity,
" virtue, and harmony." This may be true. Still
I should hesitate to say, that the man who can only
see that a cow and a camel have each four stomachs,
is necessarily a more virtuous character than the
man who can only see that one has a crooked back,
whilst the other has a straight one.

I should rather say that each faculty is equally
valuable, and that neither should preponderate.

Mr. Gamgee and his class seem to be quite ignorant that " disease" is only a word used for convenience, to represent concisely the different dis-
organizations of animal machinery. When the
boiler of a steam-engine is cracked, we do not say
it is afflicted with that terrible and malignant disease,
" enteritis," but we acknowledge we have managed
it badly, we mend it up and take better care of it
for the future. But these people seem to think
diseases to be so many actual positive entities, going
about the world like so many ramping and roaring
lions, seeking whom they may devour. Still the
word disease is of course handy to use, if only
people would remember that there is no such thing.

Mr. Gamgee seems to think that rinderpest is
perpetually bubbling up out of the ground in some
remote corner of the Russian Steppes. Dr. Unter-

berger, one of the members of the International Veterinary Congress, thought that the Siberian pest is caused by subterranean waters. Many really scientific medical men assert that cholera comes from the mud of the Ganges, and Oriental plague from the mud of the Nile. Now I don't believe one word of all this. These theories really mean nothing but this: " I am too stupid to see that men are so ignorant, dirty, covetous, vicious, ambitious, luxurious, and indolent, that they prepare themselves to contract these diseases; and if there is anything in the matter besides this, I have too much pride and vanity to confess, that I do not understand what it is." Take gout as another instance. No doubt there is sometimes hereditary predisposition; for the sins of the fathers, undoubtedly, visit the children unto the third and fourth generation. There is no escape from that law of nature. Still in the majority of cases, the explanation is more simple. But people are for some reason, scrupulous about saying, " I have been eating and drinking too much;" or, " I have been indolent and luxurious;" so they say, " I inherit the disease from my poor dear great grandfather, who was a martyr," &c.

I say I do not believe that God manufactures mud on the Nile and Ganges, in order to send forth emanations from it to decimate the creatures he has made. Probably the question is incapable of proof one way or another. Still the ingenuity of man is great. Perhaps we shall hear one of these days of experiments being made by some illustrious Parisian surgeon "in corpore vili." Perhaps we shall be hearing of a dog being shut up in a box with a table spoonful of Ganges' mud and found dead in the morning. Logic and ingenuity will prove anything. Does a question hold out? Double the ingenuity. Does it hold out still? Double it again, and the thing is done. Still as I say I shall not believe it. On the contrary, I believe that God is a benignant not a malignant being. I know well that the evil and misery there is in the world, from which there is no escape, must not be ignored. A blind optimism is as false and foolish as the opposite mistake. This evil and misery no doubt can never be explained. But it is only a small part of the whole. Still, though it cannot be explained, some partially explanatory observations can be made about it. For instance, in the animal world, almost the only natural evil is famine and its consequent diseases. But famine could only be impossible by food being so plentiful that animals would no longer have to work for their living, and the diseases consequent on this state of things would, as things are constituted, be so awful and destructive, that animal life would shortly cease to exist upon the globe. So with men, if they were not kept in some degree of order by plagues and diseases following vice, dirt and indolence, the race would soon become extinct.

Again, with regard to the miseries caused by wars

and fightings amongst men. These arise from human enthusiasms and warmth of interest and feeling. But strife could cease only by the substitution for these things of a cold utilitarianism, that is, of death. I do not deny one ideal alternative; that is, the substitution of a real Christian enthusiasm. But that, as applied to mankind in the mass, is Utopianism in this world, whatever may be the case in other planets, or at some very future day in this.

Thus we see, that though we cannot understand why the world and its contents are constituted as they are, still, being so constituted, the present miseries could only cease to be, by others infinitely worse taking their place, even to the annihilation of organized life upon the globe.

Now I have been much struck with the fact, that all through Mr. Gamgee's 859 pages, the truth that men have any power whatever towards preventing disease either in themselves or animals, is altogether ignored. I only observed one exception, where a Dr. Rawitsch said, "the indolence of the Russian "peasant is such that though he overworks his "horses, he does not trouble himself at all about "their feeding. It often happens that they die of "starvation and want of food. Their carcases lie "unburied."

Vast numbers of evils are really under control of men that are generally supposed not to be. For instance, the destruction of life at Pompeii and Herculaneum. If men had used the intelligence God

gave them, instead of consulting only their craving
for riches and luxury, these towns would never have
been built.

About contagion some say the question is of no
consequence; that all we want is to get rid of the
plague, and that the question of how it came is no
matter. I say it is of great consequence. I say
that disease in cattle, if right notions prevailed,
might in future be almost entirely prevented. My
moral in all I have said, is, that good food, cleanli-
ness, light, and open field management, would save
us from a repetition of this disease for all time. The
moral of Mr. Gamgee's book is, " They that are
filthy, let them be filthy still;" for that all the diseases mentioned therein are inevitable dispensations
of Providence. I preach food and air. Mr. Gamgee
preaches slaughter.

I recommend farmers to read the latter part of
Dr. Lyon Playfair's little book about the Plague. It
is most excellent. No doubt he adopts many of
the veterinary delusions. But he could hardly help
doing so, because all the evidence he had to go by
came through the profession. And evidence always
means what the giver of it wishes or has been taught
to be true, and therefore what he really believes to
be so. Dr. Playfair says, that the rapid growth of
the disease, is owing to our gross neglect of sanitary
laws; and that what is wanted, is by good management to make our animals proof against it as we
have already done with men in the case of Oriental
plague.

E

He also says that when well-managed cattle (are any well managed?) that when well-managed cattle get the disease, it only means that when one man's house is on fire, the next one is not safe. About the Revel story, he says it has not a leg to stand on.

In the preceding pages I have endeavoured to express my opinion about the doctrines of Mr. Gamgee and his fraternity, clearly and decidedly. But when I consider, that by their ignorant treatment,* and blood-thirsty counsels, they have caused a loss to the community of a very great many thousand pounds, I do not think I have spoken too strongly.

About Mr. Gamgee himself, I can fancy that some who may read what I have written, will think I entertain little respect for him. But they would think wrong. I feel much respect for him. It is only his logical faculty† I do not think much

---

* The most fatal part of this treatment is the depriving diseased animals of light and cold fresh air. If a diseased beast is shut up in a dark place in a temperature above 40°, no power on earth can save it, unless the case has originally been a very mild one. Then they sometimes scramble through. The drugging, if tolerably mild, probably does little harm and no good.

† At the end of Mr. Gamgee's book, he endeavours to show that Rinderpest is probably caused by minute animalculæ (entozoa) in the blood, because they can be discovered by a microscope after the animal is dead. This I call as good an illustration of " the cart before the horse" as one often meets with. The presence of animalculæ no doubt always accompanies death and decomposition, and even disease, which is often only partial death and decomposition.

of; and that is comparatively a very insignificant part of a man. Mr. Gamgee has shown great energy and great zeal, and he evidently works hard to do his duty in that state of life unto which it has pleased God to call him. Though much opposed, and sometimes considerably ridiculed, he has always treated his opponents with courteous good feeling. In October the *Times* had a leading article, holding up the irrepressible Mr. Gamgee (as it called him) to the ridicule of all its readers in all quarters of the globe. Two days afterwards appeared a letter in the *Times* from Mr. Gamgee, expressed with the utmost courtesy, and not showing the smallest particle of irritation. The qualities Mr. Gamgee thus shows himself to possess, are worth all the "logical faculty" in the world. Therefore it is that I feel much respect for Mr. Gamgee.

The article which I have just mentioned was most true and excellent, but since then, the *Times*, staunch to its principles, and true as steel to its name, has turned round once or twice in the matter. And not a word can be said against this. All well-conducted weathercocks have done just the same thing. We read in the article I am alluding to:—" The theory of importation from " abroad is one which in former times was applied " to all plagues whatever, but it has now in all " other cases been deliberately exploded. What is " said of Rinderpest was said in as many words of

" Egyptian plague. The plague used to visit us
" periodically, and was invariably ascribed to a
" shipload of infected substances. Yet we never
" get the plague imported now. That the Rinder-
" pest of a century ago should be traced to foreign
" importation was perfectly natural, for nobody
" then believed in any other origin of plagues.
" This view is so conformable to our prejudices and
" old notions; so many matters are settled, so many
" obligations dispensed with by looking at this
" plague as a spark thrown among us, to be stamped
" out with the foot, that it ought to be looked at
" with the greatest suspicion." Furthermore, we
read in the same article:—"This disease, notwith-
" standing its outlandish name, is simply ' typhus,'
" by which name indeed it is always called in
" France."

The plague has appeared in Ireland. The regulations to prevent intercourse with England have
been most stringent, so I am curious to see how the
profession will account for it. Probably they will
say that the east wind that has prevailed lately has
carried the contagium. Or we shall be told of some
beast on the opposite coast, in the delirious stage of
rinderpest, and, therefore, not accountable for its
actions, plunging into the Irish Channel, and having
been last seen swimming in the direction of the
sister island. We shall then be probably reminded
of the incredible distances animals have been known
to swim, and how stags have actually performed the

same feat in the very same place. Or, perhaps, we shall be told how some drover was seen near this newly infected spot in Ireland immediately previous to the breaking out of the disease, and how this very man was known to have been only six months previously at a farm in Cheshire where the rinderpest was raging, he having during all the time never once changed his clothes, partly from the pressure of his professional duties, partly from the known disinclination of the class to all processes of the sort, and partly from having none to change; or, if none of these will do, they will say it is not rinderpest at all, like Professor Symonds about Baron Rothschild's cattle; and I do not wish to hint a word against his absolute honesty or theirs. I only mean to say that "the wish is father to the thought," as we all know or ought to know, it always is. However, these are mere idle guesses. Most likely we shall be told something far more absurd than any of them, for the imagination of man is never powerful enough to approach the wonders of reality.

I have no doubt that veterinary surgeons are doing their duty to the best of their power and knowledge; but their opinions they take (and it is quite right they should do so) from the heads of their profession. In Oliver Twist, the butler, Mr. Giles, is examined on the subject of a burglary. Then comes the "boy Brittles," turn; the "boy Brittles," as we are told, not being actually a boy, but only called so from long habit.

"Now, Brittles, I presume you know the nature of an oath; pray, are you a Protestant?"

"Yes, sir—no—that is," said Brittles, in terrible trepidation—"God bless me—I'm the same as Mr. Giles."

So it is I find with the veterinary profession.

"What is your opinion, Mr. ——, about the cattle plague?"

"I think, sir—that is, I believe—I mean to say, my opinion—God bless me, it's the same as Mr. Gamgee's."

Thus, if Mr. Gamgee's opinions are erroneous, those of the whole profession must be erroneous also. But my endeavour in these pages has been to show that Mr. Gamgee's opinions are erroneous. At any rate, I have tried to make it clear that I think them so; and if I have failed in doing so, I think it is not my fault.

It may be said that all the authorities, together with public opinion, are on the other side. True; one hundred years ago a few, but very few, people disbelieved in witchcraft. All the authorities, from Addison downwards, together with public opinion, were on the other side.*

* I need hardly say that this reference to the old belief in witchcraft is not an argument to prove the truth of what I say, for it might equally well be used to prove any absurdity whatever—that the moon is made of green cheese, or that rinderpest comes only from Russia, or anything. No; I only make it to show that an opinion, being an opinion of the public, proves nothing whatever as to its truth.

The world, according to some, is many millions of
years old. At the very lowest computation ever made
(that of the learned Archbishop Usher) it is 5870
years old. For 5770 years of this period almost
everybody believed in witchcraft. Has the human
race all at once become so very wise, when, till so
lately, it was so very foolish. I must say I can see
no reason whatever to think it has. Forms of folly
and superstition change. Human nature remains
much the same. Of course, I know that I must not
expect assent to this view. It is contrary to the
constitution of things for people to disbelieve in their
age having arrived at true knowledge, or to see that
the knowledge of one age is the ignorance of the
next. As Emerson says, before Napoleon's time
everybody believed that the art of war had been
carried to the utmost perfection to which it was
capable of reaching. Probably the delusion is a
valuable provision of nature. For upon most people
the truth would no doubt have a disheartening
effect—the truth, namely, that knowledge is but the
knowledge of to-day, to be replaced by the knowledge of to-morrow.

The foregoing pages I have written solely with a
view to the improvement of the veterinary profession.
My only object has been their good. I could not
bear to see such a large class of well-meaning and
otherwise respectable persons sunk so deep in ignorance, so I determined to devote some of my time to
their instruction. Of course, I know there can be

no visible result. That is not the way things work in this world. Words are spoken, and seem to fall dead ; still, if they are words of wisdom, they must tell, and in fifty years ideas are not where they would have been if they had not been spoken. Disease in animals is principally a pocket affair, and therefore, only of limited importance. For this reason it cannot be wondered at that the generality of people having something better to do than to think much about the matter, should take their notions from those whose business it is, namely, the veterinary surgeons. Neither can it be wondered at that these men, to save themselves the futility of trying to think about the matter, should adopt the notions entertained by the heads of their profession. In fact, they are quite right to do so; but the result of all this is the predicament our ideas upon these subjects are now in.

I have perused the third and last Report of the Commissioners appointed to investigate matters connected with the Cattle Plague. Interesting as it is, I was rather disappointed to find the instruction contained in it lessened to me by my having already read it before ; first in various letters communicated many months ago, by Professors Gamgee and Symonds to the periodical press, and afterwards in the published accounts (which I have lately had the pleasure of perusing) of the proceedings of the Grand International Veterinary Congresses at Hamburg and Vienna. The only thing in this third report that I had not read before is the positive assertion that no non-ruminant animal can take the disease. Now this assertion has no truth in it whatever. Rabbits have been put into a close box where diseased beasts have been, and every one of them has caught the complaint, and died. I know this to be true, because it was a relation of my own who tried the experiment. This third report repeats the old assertion we have seen so often, that the disease occurs most where there is greatest traffic, or that it follows the line of traffic, as they phrase it. An old riddle asks, " Why do white sheep eat more than black ones ?" Then come the solemn guesses, such as that "black absorbs caloric, and renders less nutriment necessary," &c., &c. Then comes the real answer, " Because there are more of them." So we may put the above into the form of a riddle, and ask, " Why is there more plague in those places where traffic in cattle is greatest, than in others ?" Then come the solemn guesses, such as, " That the disease being only produced by poisonous emanations that bubble out of the ground in Russia, and are carried west by steppe cattle, therefore," &c., &c. Then comes the real answer, " Because there are more of them."

In reading these reports I have wondered a good deal how it is that Mr. Lowe, who is I believe on the Commission, does not keep the logic in better order ; but I suppose the answer is, " What is one among so many ?"

The Commissioners mention the decrease of the disease since the slaughtering as a proof of the advantages of it. .They say nothing about the observations in their previous report, that in spring these diseases always diminish and often completely die away in the summer. Mr. Gamgee said that two months of slaughter would annihilate the plague. There have been more than two months' slaughter, and there are about two thousand fresh cases every week, notwithstanding spring to help. Then comes what the Commissioners seem to think a grand conclusion, of the greatest importance, that they have quite established, namely, that the plague is caused by a specific poison. To be sure, as they allow, chemical tests fail to discover the poison, and a microscope so powerful as to make a child three feet high look as big as Mont Blanc failed also. Now, what is all this but saying with much pomposity, in many words, what none have ever doubted, namely, that some diseases are catching, but how or why nobody knows. Then we have a long story about how a drop of diseased blood will infallibly communicate the disease by inoculation. I wonder whether the Commissioners really believe that this is only true in the case of this Rinderpest, and not equally true of every kind of malignant fevers, or, as I think they used to be called, putrid fevers; that it is not equally true of famine fevers, jail fevers, Oriental plague, small pox, ship fevers, camp fevers, and every form and variety of those typhus diseases that have always been apt to follow over-crowding, under-feeding, foul air, and unnatural modes of living.

Then comes the assertion, that the whole course of the disease is seven days, which I am sorry to. say is nonsense. In one case it is two days, in another three, in another ten, and in fact there is no rule whatever. All depends on the degree of malignity. One case I knew where a beast had it one day and was quite well the third. All the others in the same yard had it and were shot. The one I speak of shivered like the others for twenty-four hours, but was well before they had time to fetch the executioner. The fact is, there are all degrees, from a slight shivering cold to most malignant plague.

Then comes the great discovery, first announced as we are told by Professor Gamgee, that when the complaint is coming on the temperature of the body rises two or three degrees. Of course it does, as in every other case and kind of violent fever under the sun. Fever means internal inflammation, and internal inflammation means abnormal internal heat.

Aberdeenshire, they say proves the efficacy of stamping out, because the number of cases in that county have been only 0·27 per cent. But there are other counties where the number of cases have been as few without stamping out; nothing is said about them. These people evidently think that one swallow *does* make a summer. There is no doubt that if Commissioners are appointed they must go by the evidence and opinions they hear. Now, when disease in animals is the question these opinions must be the opinions of veterinary surgeons. Few people seem to know the uncertainty of all evidence and of all opinion, or how invariably the evidence and opinion of men (even with absolutely honest intentions) take their whole colouring from their wishes, feelings and habits of thought, as modified by education and circumstances; so that Herbert Spencer lays it down as a law, that the world is ruled not by intellectual processes, but by feelings, wishes, emotions, desires and character.

CHAPTER V.

OUR LEGISLATION IN THE MATTER—MR. J. S.
MILL—MR. HUGHES—UTILITARIANISM—REFORM
—METAPHYSICS—MR. GLADSTONE.

I WONDER whether there ever was such a muddle
before. No doubt it was a puzzling case; for we
all know that in a free country those who govern
ought to be governed by those who are governed;
but "quot homines tot sententiæ;" so what were
they to do? There is one thing very satisfactory,
and that is the way the practical working of our
mode of government answers to the theory of it.
The theory says that the legislation should represent
and echo the opinion of the public. Now about
the cattle plague the opinion of the public was
simply confusion; and the legislation has been confusion worse confounded.

By the bye, what a curious idea that is of
modern times, that the rulers of a nation should be
ruled by the people they rule; that the legislators,
supposed to be elected for their superiority in wisdom and experience, should, instead of consulting
that wisdom and experience, consult only the pre-
sumed inferior wisdom and experience of those who
selected them; that the educated should be governed by the uneducated, and the wisdom and

reason of cultured men by the passion, folly, and credulity of the uncultured. However, perhaps it is all right. Perhaps we want a little moderate anarchy to give us a lesson. To be sure there is a lesson on the other side of the Atlantic. In the disunited States, the wise, the civilised, and the cultivated, are shut out from all part in politics, and the consequence is just what one would expect. How the revolution that is now going on there will end no one can say. Probably the disunited may become for a time united again, and all things gradually settled. But probably on the other hand, we shall be hearing some day of the assassination of President Johnson; at least he says so himself. Still America is hardly a lesson for us, for the circumstances are so different. The country is so vast that there is room for every body, wise and foolish, Mormons, Shakers, Philosophers, and Rowdies; therefore, little government is wanted. . Birds of a feather flock together without incommoding the community. The Mormons at the Salt Lake, the Shakers in some favourable spot for shaking, the Philosophers at Boston, where they can transcendentalise, with no one to disturb them; and the Rowdies in the far West, where they can swear and gamble, and bully, and murder in peace and quietness.

Ecclesiasticus says—" The carpenter laboureth " day and night, the smith worketh at the anvil, the " potter fashioneth the clay. All these trust to

" their hands. Without these cannot a city be
" inhabited; but they shall not be sought for in
" public council, nor sit high in the congregation,
" they shall not sit on the judges' seat, nor under-
" stand the judgment. They cannot declare justice
" and judgment." I wonder which is right, this
old Hebrew or Mr. Bright.* In theory it seems
desirable that the working class, as well as every
other large class, should elect a few of their number
to enunciate their little thoughts in Parliament.
For no doubt, in any country the opinions that
there are of every sort and size, had better be
known if possible. But how is it possible. How
is it possible to get five millions or so working-men
to elect suitable representatives. Take a navvy, for
instance, working on a railway. Who can doubt
what he would say if he were canvassed for his vote
for some fitting man. Who can doubt that he
would say—" I don't care a ——, I'll vote for my
" mate Jem. He's the fust chap in England, and
" can sing as good a song as any on 'em." And
quite right he would be to vote for the " fust" man
he knows. To arrive at correct opinions about
these subjects, the first requisite is knowledge of

* I found one day in Hazlitt's writings a curious catalogue
of the qualities he considered necessary to make a first-rate
popular orator. These were "force," "self-confidence,"
"want of refinement," "shallowness of thought," "absence
of originality," "being neither superior nor inferior to his
audience," "passion" and "clearness."

human nature. That is to say, knowledge of the nature of the men it is proposed to enfranchise. I myself have the highest opinion of them as a class. I suspect the uncultivated are as a rule better hearted and more spontaneously unselfish than the cultivated. But the question is as to their fitness for legislation. The leaders of the radical party are some of them men of very great talent, as every one knows. But they often seem to me singularly ignorant of human nature. Take, for instance, Mr. J. S. Mill. What can he know of human nature? Human nature means love and hate, and passion and violence, and geniality and impulsiveness, and heartiness and joviality, and good fellowship and illogical wrong-headedness. Now, what can an amiable metaphysician, who disbelieves in the existence of matter, and thinks women voting, a likely sort of thing to add to the harmony and peace of families —what, I say, can such a man possibly know about such things as these?* Most men who are at all addicted to thinking, find themselves tempted occasionally to disbelieve in the existence of matter, as in fact of every thing, and any thing. They know by expe-

---

* Hazlitt says, one objection to metaphysics is, that the study of them tends to destroy geniality, warmth, and enthusiasm. One day in an argument his antagonist struck him. A third friend who was present immediately interfered to prevent a fight, or retaliation in any way. But this was quite unnecessary. Hazlitt had no notion of the kind, all he said was, " I am a metaphysician, nothing but an idea hurts me."

rience, that a man who kills his whole nature, except the logical and reasoning faculty that is in him, that is to say, who ceases to be a man, and becomes instead only a piece of reasoning machinery —they know, I say, that such a thing must in the end come to unbelief more or less entire. Now, the utmost possible extreme of the principle of want of faith is disbelieving in every thing except what it is absolutely impossible to disbelieve in ; and the only things it is absolutely impossible to disbelieve in, are our own sensations. Now, I must say (though I cannot prove it, for it is not capable of proof, it is only a matter of faith) I must say, I cannot help thinking, that when I see a brick wall God intended me to believe in the existence (apart from my own or any one else's sensations) of that brick wall, whatever that existence may be ; and therefore I do believe in its existence.

Rigorous logical thought no doubt shows the contrary. So much the worse then for rigorous logical thought, and for the machine that turns it out. The real question is, what a *man* thinks about it. To the colour-blind, red and blue are meaningless. To the man that has no music in his soul, harmony is only a word. To the enthusiast, logic is nothing. To the logician, enthusiasm is folly. Yet all these think their conclusions right. Now, I say none have a chance of coming to right conclusions who do not possess all the faculties of the human soul and body in full perfection. If a mere logician says to me,

F

" I must be right, my arguments are so admirable."
I should say to him, "Your arguments depend
" upon the constitution of your mind: how do you
" know your constitution of mind is the right one;
" besides, bring me a greater logician than you,
" and all your fine arguments will be knocked over
" like ninepins. And so on, *ad infinitum.*"

Any man who, by a life of abstract thought, has
killed his whole nature, except the logical faculty
that is in him, must necessarily disbelieve everything,
for he has left himself nothing to believe with. The
logical faculty useful, and indeed indispensable as it
is when kept in its proper place and subordination
to the higher powers, is by itself merely critical,
negative, and destructive. The result of living wholly
and solely on the fruit of the tree of knowledge, is
death—or utilitarianism; and that is in fact only
another name for the same thing. I say, a life of
mere abstract thought is death, that is, spiritually ;
but it also is materially ; for abstract thought in its
perfection is trance, and trance is stoppage of the
breath and of the circulation, and stoppage of the
breath and of the circulation is death.

I say, there is no line to be drawn between abstract thought by itself and trance, and the produc-
tions people worship as works of vast intellect are
often, wonderful as they may be, little else than the
fantastic and morbid dreams of a lifeless and soul-
less trance or semi-death. No doubt abstract
thought does good service sometimes. Some men

seem born for nothing else ; besides, the world is carried on by division of labour. Still, such men ought to know their place. Goethe says, " a man is of importance not .in proportion to the works he leaves behind ; but in so far as he is actively happy and leads others to be so." This is probably very true. The greatest man is the man with most happiness, not with most reasoning power, still less with most what he calls knowledge. But then the happiness must be the real article, not mere pleasure enjoying. " He that increaseth in knowledge increaseth in sorrow." That means, he kills by thinking and doing nothing else, all the loving and hating and feeling and enjoying part of his nature. " Rejoice always, and again I say, rejoice." Again Goethe says, " that man loses Paradise by striving after knowledge is true of all time ; but perhaps especially true of the present."

The author of " Ecce Homo" says, that no heart is pure that is not warm ; and that no virtue is safe that is not enthusiastic. He also says, that absorbing mental activity blunts those feelings in which the life of virtue resides.

Frederick Robertson was told in Germany, that all the metaphysicians in that country were men of bad private character ; but this was I should think a very considerable exaggeration. About disbelieving in matter, that is to say, in every thing but one's own sensations, which is what abstract thought comes to at last, I will only say a few more words.

The propounders of this philosophy say, that because we cannot prove anything to exist except our own sensations, therefore, nothing else does exist. But this is no proof, it is only a possibility or probability at most. On the other hand men are so made, that in spite of logic and reasoning, they *must* believe in something else. This makes it probable that there is something else. So the question comes to this. Which of these probabilities is the greatest? There is no doubt, whatever, what every one in the world would say about it, except an exceedingly small minority. Now this minority may of course be right and the majority wrong. But now let us consider what kind of people these are, who form this minority; or rather let us state this, for their habits and peculiarities are well known. In the first place then, they pass a large part of their lives* doubled up in a crooked and grotesque posture, by the edge of a table with a goose's feather in their hands. Sometimes however they walk. But when they do it is not like other people. For they shamble with their heads bent down, and their eyes fixed in a strange vacant manner upon their own boots. And this is the more extraordinary, inasmuch as their boots are usually very ill made, and altogether without beauty. Then they never laugh, they never cry, and they very rarely smile. Their muscles are

---

* German philosophers have been known to sit at their desk for three days and three nights without moving.

of the consistency of batter pudding, their flesh, to
say the least of it, flabby; their complexion colour-
less, and their stomachs as a rule deranged.
Finally, they are dead to the interests, amusements,
and occupations of their fellow creatures. They
delight not in the things men seem intended to
delight in, and they mourn not at the things men
seem intended to mourn at. Now what must we
think of such people. Men with the savage qualities
of brutes are called inhuman. So these strange
beings without the human qualities of men, should
they not be called non-human. And if non-human,
what can be the value of their opinions about human
matters?

Then about Utilitarians. They think they
account for every thing about right and wrong, by
" greatest happiness to greatest numbers" as text,
and numerous, and I regret to say, dry sermons
upon it. But how about the passion for good and
right that all people ought to have, that many undoubtedly do have, and that great numbers have in a
more or less diluted form. Where did this come
from? Why do they not tell us that. Perhaps,
because they do not know what it means; because
they themselves have none of it. Now the accusation of not having a conscience is rather a serious
one. And yet an observer of human nature, cannot
doubt that many men have none to speak of. And
these are often very well behaved and reasoning
clever people. They are much too passionless,

prudent, and calculating, ever to do anything tangibly wrong. And they go through life generally in a most exemplary manner. Sometimes, however, a temptation comes in their way they are open to. About money most likely. Then they murder a little perhaps, or forge, or set a house on fire. It is all one. But if murder, it is done with the utmost kindness and good humour. And when they are hanged a few weeks afterwards, they no doubt think it a foolish and unreasonable penalty for the transgression of a mere arbitrarily made law of society. For if there are no such things as absolute and eternal right and wrong, any laws about conduct must be mere conventional contrivances for the convenience of society.

Of course all this that I have written above about utilitarians and metaphysicians are only general observations. They cannot be intended for any individual or individuals of either of these classes, for I am happy to say I know none. Besides, if I did, I should consider it most unjustifiable saying such things of them. "Who am I that I should judge another man's servant?"

Perhaps I ought to apologise for this digression about metaphysics, but I shall not; for after all we have still liberty in England to write what we think, however long this liberty may last. On the one hand, we are not expatriated for doing so like Victor Hugo; and, on the other hand, we have not yet come to lynching;

. . . . "and then they send all roun'
To see if there's a feather bed that's borryable in the town."

So we read in "The Biglow Papers;" the feathers
of course being wanted to apply with the addition
of tar to the person of an utterer of unpopular
opinions. But, as I say, we have not come to this
yet. As to metaphysics, of course they are neither
popular nor unpopular, most people being much too
wise to waste their time about the matter. " When
" the person who speaks," says Voltaire, "does not
" understand what he is saying, and the person who
" is spoken to does not understand what is said to
" him,—voila la metaphysique."

Montaigne says he feels inclined to hate all popular governments, whenever he calls to mind their
tendency to give way to cruel and childish superstitions; and then he tells the story of the Athenians,
in consequence of popular clamour, putting to death
their captains, because they followed up a battle by
pursuing the Lacedemonians instead of stopping behind to bury the dead. To be sure they gained and
completed a great victory; but that was a trifle compared to satisfying the popular superstition by
performing the customary ceremonies over the bodies
of the slain. Now, uncultivated human nature is
the same in all ages; of course, superstition in these
times takes different forms; but there it is just the
same. In the matter of the Cattle Plague, as we
all remember, the Government held back as long as
it could, but at last popular clamour was too loud,

credulity and superstition got the better, and slaughter was proclaimed. This time only the pocket has suffered. But curious things happen in America. Why should they not in England some day? For instance, suppose the case of a Chancellor of the Exchequer, uttering in an unguarded moment some sentiments unpopular amongst the uncultivated classes; in derision, let us suppose by way of example, of spirit rapping. Are English and American natures so absolutely different that it is impossible to imagine tarring and feathering ever taking place in this country? I do not agree with Mr. Gladstone; I am sadly afraid he is one of the kind of men Archbishop Whately alluded to when he said,—"The cleverer a man is the more harm he does, unless he has wisdom to match." I am afraid he is one of those men who can prove to themselves the truth of anything whatever, of which some influence makes them wish to prove the truth. I am afraid he is always liable to be subject to any stronger mind than his own, under whose influence he may at any time come. I say stronger, not more active, for that I should think must be impossible; I also say mind, meaning the whole mind, not mere intellect. Then he has evidently got the strange idea that breathing smoky air and propinquity to tall chimneys, and to numberless red-brick houses, in some mysterious way fits a man to take part in legislation. No doubt it is an involuntary influence exercised over him by his parentage and family. He who is born of cotton, will of course have a leaning to cotton. Human nature is human nature. It is just the same in the country. He who is born of turnips, will of course have a leaning to turnips. Still one would think that a great statesman might be superior to these weaknesses. On the other hand, I admire Mr. Gladstone's talents, I appreciate the sort of elevated and elevating tone of his mind compared with some other of the leading politicians, and I respect his earnestness in what there can be no doubt he really and honestly believes to be for the good of the country. For these reasons it is that I cannot say how I should grieve to hear, at some future time, when our institutions have become more favourable for it than they are at present, of his ever having to undergo the humiliating, and if not actually painful, at any rate distressing process to which I have alluded above.

But to return to the working classes. I wish to protest against the aspersions on them, contained in the speech in Parliament of Mr. Hughes, the member for Lambeth. In this speech he gave us to understand that the dominant feeling amongst the working classes, is a selfish narrow-minded jealousy of any member of their body (as, for instance, such a man as George Stephenson) who shows any superiority to, and therefore receives higher wages than the majority; just like stupid tyrannical school boys, who bully and thrash those who are cleverer than themselves, and who get above them in the

school. Now I cannot believe so ill of them as this.
Still, Mr. Hughes knows his friends better than I
do. When Foote, the actor and buffoon, was taken
by an aristocratic acquaintance to one of his parties,
where were collected a number of his friends, he
good naturedly said, " Mr. Foote, your handkerchief
is hanging out of your pocket." " Thank you, my
Lord," said Foote, looking round suspiciously and
thrusting his handkerchief down to the bottom of
his pocket, " Thank you, my Lord, you know your
friends∙ better than I do." So Mr. Hughes ought
to be right. Now, I know very well, that the uncultivated classes are narrow-minded and short-
sighted, and therefore unmitigated protectionists.
I know they are pugnacious in disposition, and
therefore prone to war. I know that their views
about property are apt to be (to say the least of it)
different from the views generally entertained by the
cultivated classes; and, I know that in connection
with their trade unions some terribly ugly stories of
tyranny, violence, and brutality are told.* Still, as
I say, I cannot bring myself to believe that working
men are, as a class, governed by such utter selfish-
ness and narrow-minded jealousy as their friend
Mr. Hughes gives us to understand.

By the bye, I often wonder why a man who lives
in a large town should have a vote and a voice in
the representation of the country, whilst the man
who lives in a small village has none. From sta-

* See Appendix.

tistics, the only difference between them is, that the former has less health and vitality than the other, and that he drinks more gin and less beer. Now I must say, I do not see why a small (for it is only very small), why a small inferiority in health should qualify a man for having a vote, when his more vigorous fellow-countryman is without one. But it may be the gin. Still the question remains.

However, perhaps it is all right and just; but there is one thing very certain; if it is not, the country must suffer. For wrong, injustice and unwisdom must be paid for to a hair's breadth. I only repeat this truism, because, in the House of Commons, every thing in the world almost seems to be considered, except right and wrong, justice and injustice. This interest and that interest are evidently immensely studied and cared for; but how about principle?* However, I suppose, if I were in Parliament myself, I should know more about these matters. I suppose I should soon learn, that patriotism means pliability to pressure from without; and that statesmanship means subservience of the wisdom of the few to the folly of the many.

It is a very sad thing, but there can be no doubt whatever that there are in our large towns very many drunken blackguards. Again, there can be no doubt whatever, that of these men, if they can be

---

* " I don't believe in princcrple,
  But oh I du in interest."
  The pious Editor's Creed in the Biglow Papers.

called so, a £7. franchise would include great numbers. But if it does, the question I ask is this; what have the equally drunken blackguards who live in villages done that they should not possess the same privileges. Not that I see any particular advantage that would accrue, only it seems manifestly just ; and if it is just, it *must* be advantageous ; for a country must suffer for every particle of injustice or wrong committed in it, however small, utterly unable as we usually are to trace the connection. Cause *must* have its effect. All injustice and sins against proportion are terribly paid for. If a ship is lopsided, she goes to the bottom. If a man's mind is lopsided, he is a fool. If the brain of such a man is active, he may be clever or talented to any degree, but wise he cannot be. And just by so much as he is not wise, he is miserable. So, also, it is, if the mind of a country is lopsidedly represented. As to the distinction between skilled and unskilled labour, it is mere folly. George Stephenson was an unskilled labourer, so was Robert Burns. Besides, there is no such thing as unskilled labour. All kinds of labour require skill.

Of course I know the reason why large towns are represented whilst villages are not, is that the inhabitants of villages from being scattered as they are, cannot combine, and thus put pressure upon the Government. But what an astounding thing this is to say of a civilized country.

Voltaire in one of his " Romans" describes an inhabitant of the planet Jupiter taking a turn about the solar system to see what he could see. At last he came to our earth; but concluding it was not inhabited, he departed again immediately. He said, " Ce qui fait que je pense qu'il n'y a ici personné, " c'est qu'il me parait que des gens de bons sens ne " voudraient pas y demeurer." I wonder what such a being would say if he were told that the earth *is* inhabited, and that there are in it peoples of whose governments the sole principle is " pressure." That amongst these peoples the only power that is acknowledged is " combination." That the question, " Who combine?" is never asked ; but only the question, " How many combine?" That quantity is all, quality nothing. That wisdom is powerless, because the wise are few ; and that foolishness rules, because the foolish are many; so that, if amongst these peoples the thieves and malefactors would only combine in sufficient numbers, the thieves and malefactors would be the rulers and governors. If he were told all this, I think our friend from Jupiter would cut his visit even shorter than before, but from different reasons.

I am afraid I have been rather digressing, so I will return to my subject, which was our muddled legislation about the cattle plague. As I said, the legislature must have been terribly confused by the confusion of opinion that prevailed in the governing governed classes ; but no doubt as we advance towards universal suffrage, matters will become simpler. In fact, if, as Mr. Hughes says, that almost the only idea generally prevalent amongst the masses is to bring every body to their level, both from above and below, legislation will, of course, be simplified. What can be simpler than to legislate so as to please people who have only one idea amongst them? But we have not arrived at that state yet, as we all know to our cost, from recent experience. Oh, how puzzled we have been. A man came one day to ask me, as a magistrate, if he might carry on his back a litter of pigs in a poke from one part of his farm to another, because the Inspector had told him he could not legally do so. Of course I immediately referred to the Act of Parliament and Orders in Council, and looked out for " little pigs in pokes," but it was in vain. The possibility of such a case had evidently not been contemplated, so I told the man he might carry as many little pigs as ever he liked about his farm. But whether I was right, or whether the Inspector was right nobody knows, and nobody will know to the end of time.

Amongst the numerous funny things to be found in the Acts of Parliament and Orders it is difficult to decide which to select for mention.

One clause ordains that after a diseased beast has been in any box or shed, no other beast is to be put into it for thirty days. Thus, supposing a farmer has one hundred beasts and ten spare boxes ; the first ten that catch the complaint are put into them.

They die of bullets, and are buried. Another takes the disease. Where is the farmer to put it? The only thing he can do is to build another box. All the remaining eighty-nine catch it, so the farmer must build eighty-nine new boxes.

Here is another very rich Order, namely, that when the complaint has broken out at any farmer's place he is not to remove any of his cattle, even about his own farm. But in so catching a complaint, isolation is the one sensible thing that can be done.

Verily, these Acts of Parliament should be denominated "Acts for the effectual destruction of all cattle on farms and tenements."

It seems, from a speech of Lord Derby's in the House of Lords, that in some (so-called) "infected districts" cattle were starving to death for want of food, none being allowed by the Inspectors and local authorities to be carried to them.

Then that wonderful regulation about no manure to be carried out of London. I wonder what the framers of that clause intended should be done with it. Perhaps they thought holes might be dug in the ground to bury it in. I dare say it would not occur to them to consider what should be done with the earth dug out.

"Well, Paddy, what are you going to do with all that rubbish?"

"Faith, your honour, I'll just bury it in the ground."

" But what will you do with the earth you dig out ?"

" Och, there'll be time enough to think of that."

And no doubt that would be really the idea of our legislators. For, of course, the first principle of government, is to provide for the immediate emergency as the expediency of the moment requires, and leave the rest to Providence. Perhaps they thought of Abraham Lincoln's story to the deputation, that came to him in the middle of the war, to discuss the reconstruction; how before a camp meeting in a distant part of the country, there was a grand consultation as to how a certain river was to be crossed; and how "a cunning old coon of a minister" got up and said, " My friends, I have had much experience in my life, and much wandering, and voyaging, and I have come to the conclusion never to trouble myself how to cross a river till I get to it."

However, there have been Irish Acts of Parliament before this one. I think I have heard or read of an Act that once ordained : —

(1.) That Greenwich hospital was to be rebuilt.

(2.) That the new building was to be constructed out of the materials of the old one.

(3.) That the old hospital was to be made use of, and kept standing, until such time as the new one should be completed.

" Quam parvâ sapientiâ mundus gubernatur," is a very old and very hackneyed saying; but it is so only because it is so true. Nothing but truth will stand hacking. "A lie has no legs," says the proverb.

I have not seen how the authorities got over the manure bungle. They must have got over it in some way, for otherwise we should long ago have heard of a pestilence in London. Whereas I have only seen the account of one death in consequence of it. Most likely a new Order had to be issued, somewhat to the following effect.

Whereas doubts have arisen about the construction of the present Order, relative to the removal of manure from London, the Lords of Her Majesty's Privy Council do hereby order, that

" Whenever and wherever the accumulation of
" manure in any part of the Metropolis becomes such,
" that either the disposal of it is accompanied with
" inconvenience, or a pestilence becomes probable
" amongst the inhabitants of the said district, in con-
"sequence of such accumulation, or (even though no
" such probability of pestilence should exist), the
" nuisance becomes intolerable to such inhabitants ;
" the local authorities shall have power to cause
" holes to be excavated in some contiguous park, or
" square, or garden, or in such otherwise unoccupied
" piece of ground as the local authorities shall con-
" sider most suitable for the effectual burial and
" disposal of such manure; and into these holes the
" local authorities shall cause all such manure to be
" introduced either with spades or shovels, or such

" other instruments as the local authorities shall
" consider to be best suited for the purpose."

By the bye, I wonder whether our legislators
know that manure is a thing of any value or use.
It really seems as if they are not aware that ordaining that no manure be used, is in fact ordaining that
no corn be grown next season.

If the rulers of the country blunder in this way,
and fall into the confusion they have done, what
chance is there for the subalterns. None whatever,
as far as my observation goes. Inspectors, magistrates, farmers, veterinary surgeons, and police
authorities are all at cross purposes; and the number
of hard cases and unjust decisions, to judge by my
little district, must throughout the whole country
have been innumerable. To begin with, no two
veterinary surgeons can agree what is rinderpest
and what is not; and no wonder, for there is no
absolute line between it and some other forms of
internal inflammation. Not far from me a farmer
had several sick beasts. The cattle doctor came
and pronounced the disease pleuro-pneumonia, (the
veterinary name for inflammation of the lungs, in its
various forms and consequences). Then the Inspector
came, said it was rinderpest, and ordered the animals
to be slaughtered. The owner expostulated in vain,
and slaughtered they were. Then it was shown to
be inflammation of lungs sure enough. It was like
Doctor Keats, the flogging head master of Eton.
A lot of boys were sent up to him to be examined

for confirmation, but by some mistake he thought they had been sent up to be flogged, so he set to work. The boys expostulated, but the Doctor was accustomed to boys expostulating under the circumstances, so he went on without believing a word, and flogged every one of them. Then when the ceremony was over the mistake was discovered.

A few weeks ago in a little town near the coast, between Scarborough and Hull, a man was taken before the magistrates for driving some beasts from an infected district. Upon being questioned, the man more naturally than excusably said what I am sorry to say was not true. He said he had told the magistrate who had given him the license, that the farm the beasts came from was infected; whereupon these three wise men of Gotham decided it was the magistrate's fault, and, that therefore they would not fine him. The poor man got off, so in this case it was all for the best, but still this story shows what toss up sort of work these decisions are, and though this man got off when he had done wrong, the next would probably be heavily fined, when he had not done wrong. In fact, I believe it was before the same Justice Shallows, and about the same time that a poor man was fined twenty pounds for not informing the Inspector that his cow had rinderpest, when a veterinary surgeon he had consulted had told him positively it was not rinderpest.

Now I do not for a moment mean to say that there is ever any intentional injustice, but the fact is

G 2

" wisdom " and " being qualified to act as a magistrate," are not necessarily synonymous expressions, and we are apt to get a little confused now and then with all these interminable and contradictory orders.

The following is an extract of a letter from Sir George Cholmley :—" The cattle plague is not such " a plague as the laws about the cattle plague. They " are founded upon old superstitions. . . . One " of my tenants lost many beasts before Christmas. " His farm is only two fields distant from mine. " There is always constant communication between " the farms. I never tried to prevent it, but none " of mine took it. In another situation I had others " surrounded with cattle plague ; ten oxen in a " field where they were well fed, but had not the " whole winter a shed to go into, and they are all " alive and well. The laws are absurd. I have " been most disgusted by poor men and poor women " having to walk on foot for ten miles and back, in " all weathers, to get an order for a pig to be taken " across a street. I had fourteen beasts removed " ten miles inland, with an order which the district " inspector said was quite correct. About two " months afterwards I was fined for doing this by " the magistrates assembled at B—d—t—n. Upon " what grounds I do not know, as they have not " been stated to me ; but as one of these magistrates " gave me an order immediately afterwards to re- " move some beasts back again on the same road, " their decision amounts to this : that it is lawful to

" remove cattle from east to west, but not from west
" to east."

Now I know well that our guardian angels who
issue to us their Orders in Council, have much work
to do, and therefore that it would be unreasonable
to expect them to spend time and attention upon the
ruin of only a few hundreds, or even thousands of
poor farmers and cowkeepers. More especially as
they are an uninfluential class, and one little able to
affect anyone's political prospects one way or another. Still for those who have the time it is a sad
thing to see so much distress and hardship.

***

CONCLUSION.

And now what shall I say about our prayer
with our lips to God, that "He will save that provision He has in his goodness provided for our sus-
tenance," at the same time that we are destroying
this provision and burying it ten feet deep with our
hands. It is not impiety, for none is intended.
And it is not hypocrisy; for it is done in perfect
sincerity. No. It is sheer want of intelligence in
our rulers, who order this prayer to be read under
such circumstances. Want of intelligence to see
the ludicrous incongruity of the whole thing, as
also to see the want of faith we show by our actions,
at the same time that we profess it with our lips.
These inconsistencies are very common, and the

mistake usually made is to attribute that to hypocrisy which really arises from stupidity. The Indian Thug, before he strangles his victim, offers up prayers to his deity for a blessing on his work. A religious pirate, we read of, always summoned his crew on deck for prayers before committing his deeds of murder and plunder, and whenever there was a woman amongst his captives, he always compelled her immediately to walk the plank, for fear any of his ship's company should be led into sin.

"Mr. Stiggins," said Sam Weller, "what is your tap?"

"Oh, my young friend," said the shepherd, "all taps is vanities."

"Yes, yes, I know that," said Sam; "but what is your particular wanity?"

Before the evening was over Mr. Stiggins was exceedingly drunk.

"Boy," said the country primitive, who was lord of a village grocery, "has thee sanded the sugar?"

"Yes, sir."

"Has thee watered the rum?"

"Yes, sir."

"Then coom to prayers."

Now, in all such cases the common way to account for the incongruities and inconsistencies displayed, is by accusations of hypocrisy; but the wise in human nature know that in general it all

has usually much more to do with want of intelligence. But in the present matter, besides this
want of intelligence, our slaughtering, as I say,
argues want of faith in God; or rather, it means a
firm faith in the malignity of God, but none in his
benignity. It means a sincere belief that the air
God has given us to breathe is deleterious, until it
has been mixed by men with sulphurous, carbolic or
other foul vapours;* that God has created animals
to spread poison and death ; that when He has made
his creatures, He does not know how to take care of
them ; and that his ways of doing so are altogether
wrong ways. The lowest savages want intelligence to notice the good things they enjoy, but
they cannot help noticing the exceptional evil things
they suffer. So they only notice or are conscious
of evil. Thus they consciously believe in nothing
else. That is to say, they only believe in the devil.
And only believing in him, they very properly,
under the circumstances, address all their prayers to
him ; for otherwise, they of course could not pray at
all. An increase of intelligence makes men able to
understand, and therefore fit to be told that they are
sons of God, not of the devil. Have we got this

---

* Of course I do not deny the effect of these things in
places that pure air cannot reach. But there ought to be no
such places. For the rest, if carbolic vapours were good,
I believe God would have given them to us to breathe, instead
of the air he has given us. As He has not done so, I prefer
pure air.

increase yet? I have lately been having my doubts. Ennemoser, in his book about Magic and Witch-craft, says very truly—" All these superstitions " stood on the same basis, that is, the devilish. " Protestants and Catholics alike believed that the " devil possessed at least as great power as God " himself." The popular ideas about diseases and miseries seem to me even in the present day to be founded upon exactly the same basis.

June 1st, 1866.

NOTE I.

THE WORKING MAN.

TO THE EDITOR OF THE TIMES.

Sir,—A short time ago I gave you an illustration of the liberty-loving tendencies of the Trades' Unions in the instance of the Brickmakers' Union at Manchester, who, fearing the competition of a machine brickmaking company then about to be established, determined, if possible, to prevent its ever having existence. Their plan of operations was very simple, but most effective.

The Committee of the Operative Brickmakers' Union resolved that no bricks should be supplied to any contractor or builder for any work he had in hand who directly or indirectly rendered any aid to this brickmaking company in the erection of their contemplated works.

Under this embargo the most eminent firm of builders was compelled to withdraw from a contract with the Brick Company into which they had inadvertently entered, and such was the influence of this mandate that the Company could not in all this free trading city find a builder who would undertake to help them in the establishment of their works. I related how the Company were driven to substitute a wooden house for a brick one, and to get the timber and the erection framed in Liverpool; how they were compelled to have recourse to an iron funnel instead of a brick chimney; and how one builder more courageous than the rest fell under the censure of the Union because he procured for the Company the loan of a set of " shear legs," whereby they raised their funnel to its destination, and now that the Company is in a position to supply bricks better and cheaper than the hand-made they are excluded from consumption " by order of the Committee of the Brickmakers' Union."

In further illustration of that " organization of labour," which is the object of the leaders of these unions, I wish to draw your attention to the proceedings of the stone-masons.

A contest is now going on between the master masons and their men in respect to a new code of rules and regulations which came into operation on the 1st of May. The masters in this district have, as I understand, after some show of resistance, succumbed to these rules, but object to sign them, and the strike is continued for the purpose of compelling them to attach their signatures.

By far the most serious aspect of this " organization of labour" is the attempt to shorten the hours of labour. If it be true that the position of this country depends in any degree upon the success of our manufacturing and commercial operations, it appears to me that in shortening the hours of labour we are striking at the very root of that success which has hitherto attended us.

We can only hope to maintain our position as a manu-facturing nation so long as we are enabled to produce as cheaply as our competitors abroad.

This is a question by no means confined in its conse-quences to contractors and their men—the ordinary ope-rations of manufacturing industry will soon be brought under the same rule; and although " British industry" is an enduring animal, its powers, after all, are limited.

The Legislature has done much to free the industry of this country from the trammels of Protection, and from much objectionable taxation; but if the whole labour

price of the country is to be "organized," and commanded
by a few recognized leaders, no individual trade can
defend itself against these encroachments upon the liberty
of the subject, for these regulations of labour are supple-
mented by a species of terrorism and persecution with
which the well-disposed working man is powerless to
compete ; nor can I see any reason why an organization
which is competent to control and direct the working
population as to the disposal of their labour will not be
equally effective for controlling and disposing of their
votes when they are put into possession of the political
franchise.

I am, Sir, yours, &c.,

AN EMPLOYER.

I cut the above letter out of the *Times* on the 28th of
May. Now, that intelligent men like the leaders of the
(so-called) liberal party should wish to put more power
into the hands of such men as are described in this
letter, is to me very curious; but the most curious thing
of all is, that these politicians call themselves, and actually
get other people to call them, *advanced liberals*, in favour
of *progress*, at the very time they are trying to make this
retrogressive step from liberty and freedom back towards
despotism. Verily, the Yorkshire saw is a true one,
" there's nowt so queer as fawks."

 The truth is, this extreme party ought to be called "the
" retrogressionists." No doubt the extreme Tories might
be called so too, but this only means that Hegel's grand
doctrine of " the identity of contraries " is a true one.
The division of political men into two parties is absurd.
Every thing, even a stick, has two ends and a middle, that
is three parts. There are giants, dwarfs, and reasonable

sized people. There are soft good fools,* clever headed devils, and men. There are raging fanatics, dead utilitarians, and living Christians. There are rabid ranters, ridiculous ritualists, and " sober-minded." So amongst politicians there are mad conservatives, mad radicals, and sensible men; extreme tories, extreme "retrogressionists," and reasonable beings.

NOTE II.

The following two letters I published in the Yorkshire newspapers.

Sir,—Mr. Gamgee has written a book about the cattle plague. He shows how true has turned out what he always said, namely, that no specific cure for the complaint would be discovered. He is quite right. But that very small minority, the men of clear-seeing common sense, always said exactly the same thing from analogy—from the fact that there is no specific cure for the corresponding diseases amongst men. But what an astonishing idea it gives one of the power of human credulity to think that, in spite of the experience of all ages that no specific cure has ever been discovered for any malignant disease whatever either in man or beast, that in spite I say of this, immediately what we fancy a new disease makes its appearance, we all set to work with the utmost confidence to discover its cure, and profess ourselves astonished and disappointed when we do not succeed. . In his book Mr. Gamgee repeats his doctrine that the rinderpest is only taken by contagion from Russian cattle, and that the management of cattle has nothing to do with its origin. Last summer was the

* " Tanto buon che val niente," says the Italian proverb.

warmest and longest ever known within the memory of
man. I wonder whether Mr. Gamgee ever heard that
history, from Homer downwards, is full of instances of epidemics following unusual heat. Last summer there was a
plague of flics. I wonder whether Mr. Gamgee ever heard
or read how history is full of instances of epidemics following plagues of insects. This winter has been the
wettest ever known. I wonder whether Mr. Gamgee ever
read how in London, in 1799, a malignant contagious
fever followed an extraordinarily wet season. Farmers
crowd cattle together in winter. I wonder whether
Mr. Gamgee ever heard or read how, in 1757, Sir John
Pringle records that some soldiers, upon being seized with
the common fever of the season, were confined in the holds
of crowded transports, and that the malady immediately
assumed the form of the most malignant jail fever. In
winter farmers half-starve vast numbers of their cattle. I
wonder whether Mr. Gamgee ever heard of the Greek proverb, "the plague after famine." Those beasts farmers do
not starve, they overfeed for butchers. I wonder whether
Mr. Gamgee ever heard or read how Sir B. Brodie says,
that about the most unhealthy class in the community are
gentlemen's butlers who live on the fat of the land in
regions flowing with ale and porter, and that for them to
catch even a trifling malady is a most serious thing. Then
about contagion. I wonder whether Mr. Gamgee ever
heard or read how Dr. Haygarth describes attempts which
were made without success in France at the end of the last
century to infect children with small-pox in the open air;
also how although cholera has been communicated by clothes
shut up in a box; and though in the year 1577, in the
close air of a court of justice, two judges and sundry
lawyers caught jail fever from a prisoner, and died, innumerable medical writers have recorded their conviction of

the extreme rarity of infection in the open air. I say I wonder whether Mr. Gamgee has read all these things, but of course I know he has not; nor is he for a moment to be blamed for not knowing what his education and professional duties must have prevented his having opportunity of learning. Mr. Gamgee tells us in his book, that pleuro-pneumonia (inflammation of the lungs in its various forms and consequences) is never taken except by contagion from some foreign animal. What shall we be told next?

I am, Sir,
&c. &c.

Sir,—The opinion prevails that the cattle plague is carried by cats, and rats, and dogs, and birds, but there are no facts to prove that this is so. The opinion is founded solely on our fancies as to what we imagine likely to be the case. Now this is exactly what Bacon came into the world to teach mankind not to do. He came to show that opinions should be founded upon facts, not upon fancies. Has his teaching then been in vain? One is really inclined sometimes to think so. We profess not to believe in witchcraft now, but perhaps our credulity and superstitions only happen to take other forms. A great many years ago, when the black death raged, people, in their ignorance and credulity, believed, from consulting their imaginations instead of facts, that wells were poisoned by the Jews; and the Jews were slaughtered accordingly. Now we are all believing, in the same way, by consulting our imaginations instead of facts, that infection is carried by the winds, and by birds and dogs, and we are slaughtering our cattle accordingly. Where is the difference in the degree of superstition

between these two cases. I confess I cannot see any. It is believed also that men carry the disease. Now, supposed facts
have been brought forward to show this; so the error, if it be
one, should be classed under the head of illogical reasoning,
not of superstition. Disease breaks out just after a drover
has arrived at a farm. "The drover brought it," says
everybody at once. Not a word is said about the ninety
and nine cases where no disease followed a drover's arrival
at a farm. These cases passed unnoticed. Drovers, inspectors, &c., innumerable, are always going from farm to
farm. Disease is always breaking out at farm-houses innumerable, so the coincidence of disease breaking out and
one of these men coming must continually take place.
Who is to say that this has happened oftener than the
doctrine of chances would lead one to expect? Perhaps
some extraordinarily clear-seeing, logically-minded individual might be able to do so, if he knew all the facts (which
nobody can know) certainly no body of men or committee,
for bodies of men are only very moderately logical, and
only very moderately clear-seeing. The truth is, all we
can say, with reason and without superstition, is that facts
show that one beast will often, but not certainly, catch the
disease directly from another, but that there are no facts
whatever to warrant belief in any further degree either of
infection or contagion.

I am, Sir,

&c. &c.

THE END.